The Private Life of Adolf Hitler

The Intimate Notes and Diary of
EVA BRAUN

Edited by
PAUL TABOR

The Private life of Adolf Hitler
The Intimate Notes and Diary of Eva Braun

by Francis Aldor
edited by Pal Tabori

ISHI PRESS
INTERNATIONAL

The Private life of Adolf Hitler
The Intimate Notes and Diary of Eva Braun

by Francis Aldor
edited by Pal Tabori

First published in 1949

This Printing in December, 2014
by Ishi Press in New York and Tokyo
with a new introduction by Sam Sloan

ISBN 4-87187-921-6
978-4-87187-921-7

Ishi Press International
1664 Davidson Avenue, Suite 1B
Bronx NY 10453-7877
USA
1-917-507-7226
samhsloan@gmail.com

Printed in the United States of America

Introduction by Sam Sloan

The Private life of Adolf Hitler The Intimate Notes and Diary of Eva Braun

Introduction by Sam Sloan

In November 2011, I posted a list of old books about old Nazis I was planning to reprint. My idea was pure commercialism. Books in any way sympathetic to Hitler or to the Nazis are banned or illegal in Germany but can be legally published or reprinted in the USA. So, with the idea of trying to make some easy money, I proceeded to reprint some old books from the Nazi era.

My idea has not proven successful. Nobody wants or cares about these old books any more. If I sell two in a month that is a lot.

The one book on my list that I did not actually reprint was this one. The reason is I decided that the story was dubious and probably not true. If an authentic Diary of Eva Braun were really to emerge it would create a sensation. Look at the famous **Hitler's Diary scandal and hoax of 1983**. A real Diary of Eva Braun would not create the sensation of a real diary by Hitler, but it would at least make the papers.

How can I be so sure that this diary is not authentic? Of course, I cannot be sure. However, it does not answer some questions I would like to have answered about Eva Braun and would be expected to be found in a real diary.

For example, it is well known that Eva Braun was the photography assistant and model to Heinrich Hoffmann who had a photography shop in Berlin. Heinrich Hoffmann produced a large number of propaganda pieces for Hitler, some of which I have reprinted, such as "**Mit Hitler im Westen or With**

Introduction by Sam Sloan

Hitler in the West" ISBN 487187883X and **Jugend um Hitler: 120 Bilddokumente aus der Umgebung des Führers** or Youth around Hitler: 120 picture documents from the environment of the leader ISBN 4871879100.

Some of these books show pictures of Hitler patting the heads of little blond haired, blue eyed boys. I wonder where those boys are today. They are probably dead.

Eva Braun was involved in creating these picture books. What I would like to know: Is that how Hitler knew her? Did he meet her in the photography shop? Could it have happened the other way? Perhaps Hitler knew her from some other place and brought her to Heinrich Hoffmann and got her a job there.

So we look to this book for the answer to that question and find that it does not address that important question. The first entry in this diary has Eva Braun celebrating that she has spent one year with Hitler. It appears on page 60 as "**No Date Probably September 1937**". If correct, that would mean that she first met Hitler in September 1936.

However, the book Youth around Hitler: 120 picture documents from the environment of the leader or **Jugend um Hitler: 120 Bilddokumente aus der Umgebung des Führers** was published in 1934 and Eva Braun is believed to have been working for Heinrich Hoffmann from at least that time. Thus it appears that the forger of this diary has not done his research well.

Introduction by Sam Sloan

Eva Braun was born on February 6, 1912 in Munich, Germany. Hitler was born on April 20, 1889, so he was 23 years older than Eva. There are a number of questions we would like to ask about Eva Braun and they are not answered by this book. We know that Hitler had relations with many women. Most of them did not live long. Eva Braun was merely the last one. We know she was the last one because she committed suicide with him on 30 April 1945, just one day after they had gotten married at 2:00 AM on 29 April 1945. Goebbels and Bormann signed as witnesses to the marriage. That is another question we would like to have answered. Eva Braun was a young woman with everything to live for. Why would she commit suicide with a Dirty Old Man like Hitler? She could have walked away, saved her own life and become a cult leader of the **NEW Hitler Movement** or something. Why show her devotion to him by killing herself?

You are asking why would I publish this book if I have such grave doubts about it. I decided to do so now because of reviews of a new movie about to come out based on the book "***Grey Wolf: The Escape of Adolf Hitler***". This book claims that Hitler and Eva Braun did not die in 1945 but escaped to Argentina where Braun gave birth to two children by Hitler.

This point interests me. Hitler's sister Paula Wolf survived the war. There is a man still alive in England who it is claimed is a son of Hitler by one of his mistresses from England. If these people can be located and given one of the standard DNA genealogical tests, that would determine in a few minutes whether or not this story is true and whether Hitler actually survived the war.

That book contains interesting observations such as the claim that a skull of Hitler held in a museum in Russia is not of Hitler and the supposed bones of Hitler do not show signs of syphilis. Hitler is known to have contracted syphilis from a prostitute at aged 17.

The book here on page 146-147 shows Hitler being attracted to a nude 14 year old girl they had just caught in a bathtub while she was taking a bath. Eva tells Hitler to take the girl if he wants her because she is sure that Hitler is attracted to her. I am not sure what this is supposed to show or prove.

On page 100 Eva says that she has finished reading **Marie Antoinette** by **Stefan Zweig**. It so happens that I reprinted that book in 2010 just because it seemed like a good book without knowing this history. However if you search www.amazon.com you will not find it listed. The ISBN number is 4871878554. However, you will find it listed on Amazon in Germany http://www.amazon.de/Marie-Antoinette-Portrait-Average-Woman/dp/4871878554 and Amazon in Italy. http://www.amazon.it/Marie-Antoinette-Portrait-Average-Woman/dp/4871878554 There is a surprising explanation for this. After the book came out, I started receiving letters from an attorney in Switzerland who wrote that I was violating the copyright of his client Williams Verlag. I responded that this book was first published in Leipsig, Germany in 1932 as **Marie Antoinette. Bildnis eines mittleren Charakters**. All books published in Germany during or before the Nazi era had their copyrights canceled by President Franklin Roosevelt under **Trading with the Enemies Act** in 1941, so there is no copyright on that book under US law.

The lawyer representing Williams Verlag wrote back that Stefan Zweig was a Jew and an opponent of the Nazis and it was not fair to punish him because of what the Nazis had done. I responded that in any event Stefan Zweig had famously committed suicide in 1942 and thus he had no copyright.

In response, this lawyer wrote letters to all the major US book distributors threatening them with legal action if they continued to carry this book. As a result, all the US distributors dropped this book. For that reason you cannot find it on Amazon. Nevertheless, that did not seem to bother the German and Italian distributors so they continue to carry this book.

There are several other reasons why I doubt the authenticity of this book but I will leave that for the reader to decide. For one, the translation is too smooth. Anybody who has done translation work knows how difficult it is to produce an attractive, readable work out of a translation. Eva Braun was supposedly a poor German girl with limited education. How could she have produced a book like this in flowing English, even with the help of a translator?

This book was first published in London in 1949 by Francis Aldor and edited by **Pal Tabori**. All we know about Francis Aldor is he filed for Bankruptcy shortly after the publication of this book. He has only one other title to his credit. The Good Time Guide to London (1951).

Pal Tabori (or Paul Tabori) was a Science Fiction writer in Budapest, Hungary. This does not cause us to place much confidence in the accuracy of a work of historical significance.

Another book has appeared entitled "The Diary of Eva Braun" published in 2000 by Spectrum International. This appears to be the same book, as the source is the same. Luis Trenker is the source and it has the same story behind it.

The source for this manuscript is Luis Trenker. Luis Trenker was a recognized film producer. He was born **Alois Franz Trenker** on 4 October 1892 in what was then Austria but is now part of Italy. The story is that he was staying in the same hotel as Eva Braun and she handed him this manuscript in a sealed envelope. This is credible. However, he lived until 13 April 1990 when he died at age 97!! in Bolzano, South Tyrol, Italy. If he was the source of this manuscript it is surprising that he did not seem to say anything about it.

A diary of Eva Braun dated 1935 has been found and is deemed authentic. It is 22 pages long.

Introduction by Sam Sloan

Twenty-two pages of Eva Braun's 1935 diary were found after the war. It is a fascinating document of an immature young woman whose love for Hitler is unmistakably sincere. This early, authentic diary for the period from February 6 to May 28 1935 exists, both in German script and in English typescript translation, in the National Archives, Washington, D.C. Here is the entry for the first days:

"February 6, 1935

I think this must be the right day to begin this extra-special diary. I have now reached the happy age of 23. No, happy is not quite the right word. At this particular moment I am certainly not happy.

The truth is that I have rather large ideas about the importance to be attached to this day: If I had a dog I would not feel so lonely, but I suppose that is asking for too much.

Frau Schaub came as an ambassador bringing flowers and telegrams. The result is that my whole office resembles a flower shop and smells like a cemetery chapel.

I suppose I am ungrateful, but I did want to be given a dachshund. And I just don't have one. Perhaps I'll get one next year, or much later, when it will be more appropriate for a budding old maid.

What is important is not to give up hope. I should have learned to be patient by now.

Today I bought two lottery tickets,

because I had a feeling that it would be now or never--they were both blanks. So I am not going to be rich after all. Nothing at all to be done about it.

Today I was going to Zugspitze with Herta, Gretel, Ilse, and Mutti, and I should have had a wonderful time, for it is always most enjoyable when other people are enjoying themselves, too. But nothing came of it. This evening I am going to have dinner with Herta. What else can you do, when you are a little single woman of 23? So I shall end my birthday "with gluttony and drunkenness." I think this is what he would want me to do.

February 11, 1935

He came to see me, but nary a sign of a dog or a chest of drawers. He did not even ask me what I wanted for my birthday. So I bought some jewelry for myself. A necklace, earrings, and a matching ring, all for 50 marks. All very pretty, and I hope he likes it. If he doesn't, then he should choose something for me himself.

February 18, 1935

Yesterday he came quite unexpectedly, and we had a delightful evening.

The nicest thing is that he is thinking of taking me from the shop and -- but I had better not get excited about it yet -- he may give me a little house. I simply must not let myself think about it. It would be marvelous. I wouldn't have to open the door to our "beloved customers," and go on being a shopgirl. Dear God, grant that this may really happen not in some far-off time, but soon.

Poor Charly is ill and won't be able to come with me to Berlin. But perhaps that is best after all. Br. can be very rude to her sometimes, and that would make her even more unhappy.

I am so infinitely happy that he loves me so much, and I pray that it will always be like this. It won't be my fault if he ever stops loving me.

I am so terribly unhappy that I cannot write to him. These notes must serve as the receptacle of my lamentations.

He came on Saturday. Saturday evening there was the Town Ball. Frau Schwarz gave me a box, so I absolutely had to go after I had accepted. Well, I spent a few wonderfully delightful hours with him until 12 o'clock and then with his permission I spent two hours at the ball.

On Sunday he promised I could see him. I telephoned to the Osteria and left a message with Werlin to say that I was waiting to hear from him. He simply went off to Feldafing, and refused Hoffmann's invitation to coffee and dinner. I suppose there are two sides to every question.

> Perhaps he wanted to be alone with Dr. G., who was here, but he should have let me know. At Hoffmann's I felt I was sitting on hot coals, expecting him to arrive every moment.
>
> In the end we went to the railroad station, as he suddenly decided he would have to go. We were just in time to see the last lights of the train disappearing. Once again Hoffmann left the house too late, and so I couldn't even say good-by to him. Perhaps I am taking too dark a view, I hope I am, but he is not coming again for another two weeks. Until then I'll be miserable and restless. I don't know why he should be angry with me. Perhaps it is because of the ball, but he did give his permission.
>
> I am racking my brains to find out why he left without saying good-by to me.
>
> The Hoffmanns have given me a ticket for the Venetian Night this evening, but I am not going. I am much too miserable.

Note the difference in style. It reads like a real diary. This diary ends with Eva writing that as she has not heard from Hitler in three months, she is going to take 35 pills and kill herself if she does not hear from him by the end of the day. Obviously, this plan was not carried out or it was not successful.

Here is a list of old Nazi books I have reprinted, in ISBN number order. This includes books on both sides of the war plus books unrelated to the war but that happened to be published in Germany during the Nazi period:

0923891730 Inside the Third Reich
4871878791 Spandau The Secret Diaries
4871878775 Infiltration: How Heinrich Himmler Schemed to Build an SS Industrial Empire
4871878805 The Loneliest Man in the World The Inside Story of the Thirty Year Imprisonment of Rudolf Hess
4871878813 Berlin Diary
487187883X Mit Hitler im Westen or With Hitler in the West
487187897X Speeches by Hitler in 1938
4871879100 Youth around Hitler: 120 picture documents from the environment of the leader Jugend um Hitler : 120 Bilddokumente aus der Umgebung des Führers
4871879119 Der Triumph des Willens Kampf und Aufstieg Adolf Hitlers und seiner Bewegung Triumph of the Will. Fight and Ascent of Adolf Hitler and its Movement
4871879127 The Kersten Memoirs 1940-1945
4871879135 Life for Hungary: Memoirs of Admiral Nicholas Horthy Regent of Hungary
4871879143 Hitler Directs His War The Secret Records of His Daily Military Conferences
4871879194 Die Freiwilligen der Waffen - SS Idee und Opfergang by Felix Steiner
4871879208 Hitler's Strategy
4871879216 The Private Life of Adolf Hitler The Intimate Notes and Diary of Eva Braun
4871879224 Die Armee der Geächteten or The Army of the Outlaws

Introduction by Sam Sloan

4871879259 Diary of Anne Frank in Dari Persian or Farsi
4871878554 Marie Antoinette The Portrait of an Average Woman
4871878562 Conqueror of the Seas The Story of Magellan
4871878570 Amerigo A Comedy of Errors in History
4871878589 Mary Queen of Scotland and the Isles
4871877094 The Trachtenberg Speed System of Basic Mathematics

The diary ends on July 23, 1944. It reflects the failure of the **20 July plot** which was the attempt to assassinate Adolf Hitler inside his Wolf's Lair field headquarters near Rastenburg, East Prussia, on 20 July 1944. The primary purpose of the assassination attempt was to seize political control of Germany and its armed forces from the Nazi Party (including the SS) in order to obtain peace with the western Allies as soon as possible. Had the plot succeeded thousands of lives would have been saved. The way this book reached the publisher is Eva Braun was sitting in a restaurant and she saw a famous movie producer, Luis Trenker, sitting there, so she handed him this diary. It could be that she was just in the middle of writing something and seeing him, she took advantage of the moment, knowing she would not see him again.

Sam Sloan
Novato California
USA
December 14, 2014

PUBLISHER'S PREFACE

The Third Reich, as we know now, was built on a foundation of "bigger and better lies." Hitler, ignoring Lincoln's sound principles, tried to fool *all* the people *all* the time. No wonder that when the Nazi régime, which was intended to last a thousand years, crashed after twelve, the most contradictory tales emerged about the Nazi leaders, their crimes and their passions. Much of this material was deliberately manufactured by sensation-mongers; a good deal of it served political purposes. The testimony, the charges and details were so confused that sometimes even genuine disclosures became suspect, and were doubted by those who found themselves bewildered by contradictory evidence.

There was another reason for this suspicion and bewilderment. The basic disease of Nazi Germany was moral insanity. Never before in world history, not even in the declining days of the Roman Empire, had such a gang of sadists, sexual perverts and madmen gained power over a large country. The Nazis had no moral inhibitions. Their crazy standards, their lunatic conception of good and evil, were so alien to normal men that even to-day we cannot quite believe the proofs of their depravity.

This applies especially to Adolf Hitler. Except for the turgid and infamous "Mein Kampf," and his speeches, the late unlamented Fuehrer has left no record of his life and thoughts.

Of his personal traits, preferences, aberrations and ideas we have gained scattered evidence from the works of Konrad Heiden, Hermann Rauschning and others. But of his private life, especially after he came to power, practically nothing reliable has emerged.

This is the importance of the following work. As far as it is humanly possible, we have investigated the authenticity of Eva Braun's diary and—whatever degree of authenticity one may attribute to the proof of her authorship—we are convinced that here is a first-hand, detailed, shockingly frank account of Adolf Hitler's love life. It is necessarily fragmentary and scattered over several years, but even so it gives a vivid picture of the man who unleashed the untold miseries of the second world war upon mankind.

We owe the reader an explanation as to how this diary was preserved and its authenticity established.

Luis Trenker is a well-known Austrian writer, film-director and film-star who has done the Allies valuable service during the war. Starting as a mountain guide and skier, he graduated to writing books and filmscripts and became one of the favourite outdoor stars of German and Austrian films. Because of this he had many contacts with leading Nazis—whom he despised but, for reasons of his undercover work, continued to frequent. He also met Eva Braun—how and under what circumstances you will read in the first part of this book—and in the winter of 1944/45 spent some days at the same hotel in Kitzbühel, the famous Austrian skiing resort, where the Fuehrer's mistress was staying. During these days Eva Braun handed him a sealed package, requesting him to keep it for her.

In the autumn of 1945 Trenker moved to Bozen. There, going through his possessions, he came upon the sealed package again which was marked with the initials " E.B." As he knew that Eva Braun had died with Hitler in the Reich Chancellery on April 30, 1945, he took this package in December, 1945, still sealed, to a public notary in Bozen/Bolzano and opened it in his presence. They found a pile of loose pages in typescript, bearing no handwriting, and no identifying mark. But it soon became evident that these were the irregular jottings of a woman who was Adolf Hitler's mistress; and further reading revealed that they must have been written by Eva Braun.

Luis Trenker put the material at our disposal. He also provided some data for the first part of this book which deals with the other women in Hitler's life, and tries to solve the sexual riddle which he represented. In this first part we collected all the material available from first-hand sources, incorporating the testimonies of Trenker, Heiden, Rauschning and many others. The second part of the book contains Eva Braun's diary in literal translation, except for certain passages which, because of their obscenity, cannot be reproduced, together with some explanatory notes about the personalities and events to which she refers.

We believe that this is a valuable contribution to the history of Nazi Germany and to the personal portrait of its Fuehrer, Adolf Hitler.

FRANCIS ALDOR

PART ONE

THE WOMEN IN HITLER'S LIFE

1

PRACTICALLY all the books dealing with Adolf Hitler, from Konrad Heiden's great biography to Hermann Rauschning's *Conversations* contain a chapter about Hitler's relations to women. This is easy to understand, for in this intimate sphere the character of any human being is disclosed with fewer inhibitions than in any other. But it applies especially to a man who was one of the greatest political mountebanks, wearing a thousand masks but never showing his real face. "In the fight for power it was due to the votes of women that Hitler succeeded at last," Heiden says. Rauschning maintains that "the Fuehrer was discovered by women; after the first world war, various ladies smoothed the path of the still young political adventurer." The wives of some great industrialists gave him money and, during the period of the disastrous German inflation, works of art, to finance his political schemes. It was in a circle of politically-minded women that the paid propagandist became a political prophet. "These women spoiled him, increased his self-importance by the praise and admiration which they advanced, so to speak, before he had achieved anything. The adoration of women, heightened almost into a pseudo-religious ecstasy, became for him an essential stimulation in order to overcome his innate lethargy."

This might well be true. But most of these women were elderly wives or spinsters who sat in the first row at all the meetings at which Hitler spoke, and who looked at him with their moist, admiring eyes as if he were a god. They hardly belong to the chapter of Hitler's sex life; at the most, they illustrate the rôle of eroticism in modern mass propaganda. This evidence does not provide an answer to the question of whether the Fuehrer achieved any personal and private success with women.

The official biographers of Hitler keep at a respectful distance from the delicate problem of sex; and even the anti-Nazi biographies only indulge in dark and mostly meaningless hints. They only agree on one point: that Hitler was not a homosexual. Was he "normal"? We know to-day that he pretended to be a man without any private life—which was just as false and untrue as most things about him. More than one woman had committed suicide because of him. Konrad Heiden offers the theory that Hitler was neither homo—nor heterosexual, but a "*hörig,*" man, strongly bound or inhibited in his sexual life. He is certainly right when he states that all his relations with women were "dark and mysterious." It is another question whether Heiden's further conclusions were right. "His affairs," continues Heiden, "all seem to break off at a certain stage, almost without exception, and in many cases it can be recognised that it is Hitler who was deserted and not the women whom he left. One of the women who was an intimate friend of his, declared, when asked about her relations with him, that she had experienced a disappointment which left her with little respect for Hitler as a man."

Madame Tabouis, the French authoress and journalist, went even further. She declared that according to the personal experience of a lady whom she knew well, Hitler was impotent. This, in view of Eva Braun's diary, has been proved utterly and completely wrong.

It is, of course, impossible to give a complete chronological history of Hitler's sex life. But we can attempt to group the available evidence in such a way as to follow the main periods of his career and show what part women played in it.

2.

Erna Hoffmann, the wife of Hitler's "court photographer," declared in an interview which she gave to a French reporter and which was later reproduced in a Swiss illustrated paper, that Eva Braun was by no means Hitler's most important lady-love—only "one of his twelve mistresses." Frau Hoffmann said that Hitler had deserted Eva "just as he deserted his first mistress, the Jewish girl Johanna Wachsmann, who in 1913 had left her rich parents in order to live with Hitler on the fifth floor of the Vienna inn, 'Zur Schwarzen Katze.' Later he also deserted a nurse, then a Miss Grotzki; he also discarded a romantic Englishwoman and a frigid Polish girl. There was a peasant girl, too, among Hitler's mistresses. But he broke with her when, on the occasion of his visit to the Berlin Opera, she called out loudly to him: 'Hey, Adolf, listen!'"

The "twelve mistresses" whom Erna Hoffmann enumerates cannot be taken very seriously. Many similar legends were created by Hitler's predilection for pretty, fair-haired girls. He liked to seat them around him at table, petting them and permitting himself small familiarities. Rauschning describes such scenes, which he himself witnessed. It is true that his early party comrades had often reproached him with "too much dalliance"; but Gottfried Feder, the early theoretician of the Nazi Party, admitted that he had the right to refresh himself " in a circle of lovely women."

Hanisch, a friend of Hitler's youth, has a different opinion. In a pamphlet which he published about Hitler's years in Vienna, he ascribed Adolf's timidity to the fact that he never had any success with women. One of his early fellow-Nazis, Heinrich Dolle, accused him of " sitting around too much with females in the *Fledermaus Bar*," which wasn't good for his soul. About this time Hitler bragged a good deal about his love affairs, referred to himself as " the king of Munich," and tried to create the impression that he was a veritable Casanova. One of the men who knew him best maintained that he was incapable of genuine affection. His love of feminine company was only " a public comedy." The whole dark, demonic nature of this man was expressed in the most elementary relationships. He remained right to the end the " sensitive adolescent," deeply hurt in his vanity.

Before he came to power, Munich society was rather " shocked " by his interest in Erna Hanfstaengel, sister of Ernst Hanfstaengel, who quarrelled with Hitler soon after the first triumphs of the Nazi Party and subsequently went to America. Erna was a slim, tall, proud, cool and very lady-like girl. She treated the leader of the party almost with condescension, kept him at a certain distance, and gave him little hope. But his passion was so violent that he refused to be banished from her society. The Fuehrer's friends, who were constantly afraid that he was making himself ridiculous, spread the rumour that Hitler had become engaged to a lady " of Jewish ancestry." Thereupon Hitler inserted an announcement in the *Völkischer Beobachter* which he composed himself: " Rumours have been spread about Adolf Hitler's engagement to a Jewish lady. These rumours are lies. In any case," and here he displayed typical Nazi tact by naming her " Fraulein Hanfstaengel is not of Jewish descent."

If Hitler had thought that he was making Erna Hanfstaengel's social position impossible with his clumsy stratagem, he was badly mistaken. Shortly afterwards she married the well-known surgeon, Dr. Sauerbruch, and settled with her husband in Berlin.

Later, in 1931-32, Hitler was a frequent guest in Winifred Wagner's house in Bayreuth. It was an open secret that Richard Wagner's daughter-in-law and Siegfried Wagner's widow had set the Fuehrer's heart on fire. Again, after a while, there were rumours of an engagement in the Party. But in the autumn of 1932 this friendship ended suddenly. Hitler, who had stayed so often in Haus Wahnfried, Wagner's palatial home, left Bayreuth and, without any explanation, moved with his retinue to the nearby town of Berneck. He still kept his interest in the home of the Wagner Festivals—but he no longer paid visits to Winifred Wagner. Once again an emotional relationship had been interrupted suddenly and mysteriously.

Very early in Hitler's career, about 1923, Jenny Haug was known among the party members as the Fuehrer's fiancée. Jenny's brother was Hitler's chauffeur. She was very jealous of young Mrs. Esser, wife of an early party fighter, and of other women. Hitler reassured her " jovially " and called her " a silly little creature." These old party comrades maintained: Hitler had promised Jenny that he would marry her.

Henny Hoffmann, daughter of the photographer Heinrich Hoffmann, was also Hitler's favoured companion for a short time. It was said that in her presence Hitler was always uneasy and lost his self-control. But nothing came of this, and Henny later married Baldur von Schirach, the Youth Leader of the Reich.

What strikes us in Hitler's early relations with women is his inconstancy, his changing taste and his faithlessness. " It seems," one of his biographers wrote, " that he liked to mix the respectful admiration of polished women with the cruder relationship of more earthy types." Once he flew into a rage because one of his friends pinched a waitress in his presence " on the upper leg." On the other hand his more intellectual friends frequently complained about the " impossible doxies " whom he introduced into every kind of society and on the most unsuitable occasions.

The companion of his youthful Vienna days, Hanisch, tells us that whenever they discussed women, Hitler delivered speeches full of worldly wisdom, maintaining that self-discipline was the most important of all virtues. On the other hand he also claimed that " all women could be bought or tricked into yielding." It was only necessary to be daring. " Push back your hat, show your face and then go ahead!" Yet he also hinted that easy conquests did not attract him. Hanisch describes a very early youthful experience while Hitler was in the military hospital. There was a young servant-girl there to whom Hitler was attracted, nor was she reluctant to yield her favours. Once, when she was milking a cow in the stables—it was dark and warm—something " almost happened." But in the very last moment Hitler's conscience pricked him. " You are so young!" he cried to the girl, freed himself abruptly, and rushed out, kicking over the milking bucket in his flight. He felt very satisfied with his self-restraint and decency. Was this " decency "—or was it fear?

3.

One could collect rumour and gossip, reports of fleeting adventures and unimportant liaisons by the dozen. But most reliable witnesses agree that there were only four women in Hitler's life who were important to him for various reasons and in different degrees; four women—instead of the dozen Erna Hoffman mentioned. And of these four there is only one of whom we have certain proof that she was Hitler's mistress.

The first " great passion " of his life was also a dark tragedy—though not for him. It was his love-affair with his niece, Geli Raubal. The daughter of Angela Raubal, Hitler's step-sister, Geli was a young, fair-haired, well-developed, typical peasant girl from Upper Austria. She was born in 1908; she came to Munich after the war and studied music; she also had vague ambitions of going on the stage. She lived with her Uncle " Alf " in the Brown House, Briennerstrasse. This caused gossip and indignant murmurs in the Party. In 1928 Hitler insisted that the Wurttemberg Gauleiter Munder should be expelled because he permitted himself " improper remarks " about the relationship between " Alf and Geli." In those days, it seems, Geli adored Hitler as her " big, famous uncle." It must have been a childish but unconditional admiration very welcome to the " black-haired man with the ugly forehead who kept on picking his teeth while he listened to people." He, too, became very fond of the young girl who idolised him. As the years passed, their relationship became unhealthy and oppressive. Geli's original, naive ardour turned gradually—we do not know by what stages—into a deep repugnance against Hitler who evidently could no longer restrain his passions. Geli described her uncle as " a horrid fellow " and Hitler, whenever he had one of his notorious fits of rage, abused her for a whore. He forbade her the company of other young people and once threatened her when he met her with one of her fellow-students in the street, to chastise her with the whip he always carried at that time.

Emil Maurice, of whom Hitler seems to have been especially jealous because Geli Raubal confided in him, related that " Alf " used to deliver long lectures to the girl, spiced with obscene details, about the dangers of sexual intercourse. On one occasion, when Geli stayed out till three o'clock in the morning, Hitler forced his step-sister to take her to a gynæcologist the following day. He had to know whether she was " innocent or not." While the two women were with the doctor, the Fuehrer walked excitedly up and down in front of the house. Geli, who had resisted this examination to the last moment, was dragged along to the doctor by force. She was then twenty years old. The examination proved that she was a virgin. Hitler took her home, deeply satisfied; the girl was dazed by repugnance and shock. Her uncle overwhelmed her with promises if she " behaved well " in future. On the same day he gave her a valuable ring—but in the evening he locked her in her room.

At last Angela Raubal, Geli's mother, became uneasy and demanded an answer from Hitler about his intentions. According to Emil Maurice, the Fuehrer gave an evasive reply. When the anxious mother pressed him with her questions, he began to rave, shouting that Geli was a " cunning hussy." Angela Raubal replied that a doctor had just proved her daughter's purity; whereupon Hitler screamed that Geli was a *demi-vièrge* and he could not be deceived so easily. Frau Raubal left the room and gave her daughter the long-awaited permission to visit her friends in Austria. Hitler had one of his crying-fits, screamed, trembled and collapsed.

In the autumn of 1931 the whole of Germany learned about this affair through its tragic ending. Geli Raubal wanted to leave Munich, she had long planned to go away and live in Vienna. Until this time Hitler had always been able to bully her into staying. There was a last quarrel between them in the presence of some friends. Geli stuck to her decision. Hitler rushed from the room, and down the stairs to his waiting car. Geli followed him—perhaps she felt the need for a last conciliatory word before she went away; for her intention was never to return. Standing by the car, Hitler told her once more that she was not to leave. His face had become a twisted, threatening mask. Then he drove off—to Hamburg.

" The young girl," writes Konrad Heiden, " begins a letter to him which is never finished; it contains no reference to the events of the next few hours; several other actions and utterances show her calmness and her serenity. Her mother is away in Berchtesgaden; Geli is almost alone in her uncle's home . . . Next morning she is found, shot dead; no one heard the shot in the night. It is the 18th September, 1931."

The police surgeon declared that it was suicide—a bullet through the heart, fired from above. Two days later her body, transferred to Vienna, was buried hurriedly in the Central Cemetery. Through the influence of a Roman Catholic priest, Angela Raubal contrived to have her daughter buried in consecrated ground though the Church normally denied this to suicides. She was able to do this because she could prove that suicide was practically impossible. The young girl was buried in a plot which had no head-stone nor any marking up to the year of the *Anschluss*—nothing except a cardboard shield with a black border and the words: " Here rests in eternal sleep our greatly beloved Geli. She was the sunshine of us all. Born 4.6.1908, died 18.9.1931. Raubal Family."

It was said that for the first few days after Geli Raubal's death Gregor Strasser did not leave Hitler alone for a single moment because he was afraid he would commit suicide. But this was only a camouflage. Hitler, who at that time could not visit Vienna without the permission of the Austrian government, received this permission at his urgent request in order to spend some time at Geli's grave—with the express proviso that he would not engage in any political activities. The Austrian Nazis were ordered by Munich not to take any notice of the Fuehrer's visit.

Hitler arrived by car in Vienna. It was late at night. The cemetery gates were already locked; they had to be opened especially for him. Alone in the darkness he walked to the grave. The same night he returned to Germany. A year later Ziegler, the painter, was commissioned to paint a life-size portrait of Geli Raubal from a photograph. Hitler, when presented with the portrait, burst into tears.

As dubious rumours persisted about this suicide, the Bavarian Ministry of Justice ordered an official investigation. It was discovered that there were many circumstances excluding the possibility of suicide; yet the investigation did not produce any concrete results. A highranking Bavarian civil servant who knew the case well, declared years later that it was a " dirty story." Konrad Heiden, referring to the version current at that time in the party, that Hitler shot his niece himself, declared that this was certainly untrue. Hitler was not in Munich at the time. According to Heiden another Nazi, someone very close to the Fuehrer and who later achieved a high position in the Party, was involved in the matter. The Bavarian—under Hitler later Reich—Minister of Justice who ordered the investigation and allowed it to end in failure, was Wilhelm Gürtner.

Is this all we know? In the autumn of 1945, the American army paper " Stars and Stripes," reported the discovery of documents which were said to have proved that Adolf Hitler fired the fatal shot himself. Other details were not published. It had been discovered earlier that Adolf Hitler did not spend the night of Geli Raubal's death in Hamburg, in spite of the official declaration. Psychologically the American statements seem to be credible—Eva Braun's diary gives us the answer why.

The question remains—what was the motive of this murder? A girlhood friend of Geli Raubal gave a French officer some details which were published in the " Paris Soir." According to this the young, gay, but certainly pure-minded girl who was not thinking of amorous adventures, had suffered silently for a long time from the pressing attentions of her uncle. But when she was unable to fend off his wooing, she confessed everything to her mother. Angela Raubal forced Hitler to give his word of honour not to seduce Geli. But this was only the beginning of the tragedy. After this violent scene, according to the " Soir," Hitler spared Geli physically but demanded of her a different kind of satisfaction. This raised the girl's reluctant dislike to outright repugnance. What these demands were might, perhaps, be deduced from Hitler's violent reference to Geli's condition as a " *demi-vièrge.*" Yet all this offers no explanation for the alleged murder. Was he jealous? Jealous of Geli's possible future suitors?

Later years gave plentiful proof of his mysterious jealousy and greed for possession. "His jealousy is peculiar and inhuman," Eva Braun says in her diary.

Geli Raubal's death must remain a mystery. Whether Hitler fired the fatal shot himself or drove her to suicide, he was certainly directly responsible for her early death.

4

After the Nazis came to power, Hitler became considerably more discreet in his amorous adventures. As the Chancellor of the Third Reich, he could not permit the slightest public scandal to sully the name of "Germany's saviour." As an experienced propagandist, he took good care that only those details of his private life should be made public which he considered suitable—and even these were transformed into touching anecdotes and moving legends. That these had little relation to truth naturally did not disturb the believer in "bigger and better lies."

One of the women for whom he conceived a short-lived passion was Margarete Slezak, daughter of the great Viennese singer, herself an accomplished *artiste.* But this, too, was a one-sided affair. Margarete Slezak was much too fine and independent a person to become the Fuehrer's mistress. When Hitler realised this he did not, for once, banish the reluctant lady from his circle. She became a sort of lady-in-waiting at his court. She was a gay, intelligent, and always even-tempered woman. But later, when she began to make remarks about things she disliked in the Nazi party and even attempted to change Hitler's mind, she was quickly removed from his intimate circle.

Once, as she told Luis Trenker, she asked Hitler why he was so much opposed to religious faith in his officers and soldiers. Margarete Slezak had once visited Rome where she saw a whole regiment kneeling in front of an altar being blessed by the Pope—a sight which had moved her deeply. Hitler made some evasive reply. When Miss Slezak left the Reich Chancellery, she was met by Schaub (for many years Hitler's aide-de-camp), Himmler and Sepp Dietrich. All three attacked her violently and threatened her with serious punishment if she ever dared to "interfere" in such matters. She was never again admitted to Hitler's presence.

It was Schaub, incidentally, who told Margarete Slezak about Hitler's taste in women. It was his task, among other things, to visit theatres, places of amusement and music halls, seeking suitable companions for the Fuehrer. "The Fuehrer," he told Fräulein Slezak, "has no use for actresses; he dislikes women who have an intellect—he needs young, uncomplicated girls; the sort you find in music halls are about his type."

In the winter of 1932-33 Hitler spent a good deal of his leisure in the house of a retired admiral whose niece seemed to interest him. He took her on motor excursions—always accompanied by her mother for she was not allowed to go out without a suitable chaperon. Once again there were rumours of an engagement—but the pretty, twenty-year-old girl eloped with a lieutenant who was only three years her senior and cabled after a week from Hamburg that they had got married.

Another young girl, daughter of a Berlin scientist in whose house Hitler and Goebbels were frequent guests during the "fighting years" of the party, seems to have fascinated him for a while in the spring of 1933. Konrad Heiden relates this episode in the following passage: "The lady of the house addressed Hitler as 'Herr Adolf.' He had to listen to her reproaches about the Jewish boycotts which were strongly attacked in this cultured house; then Goebbels intervened and said with pretended regret: 'But my dear lady, *I* am the one to blame!' The only daughter of the house, a gay young girl, found pleasure in leading the most famous German of her age by the nose for a little while. He accompanied her with amazing patience on her motor-tours—but that was all."

Now that he had become more selective and cautious, he picked the women he liked by proxy and had them visit him in secret. But one must be cautious in enumerating all the women with whom Hitler was reputed to have had an intimate relationship. Eva Braun's diary gives us a definite indication why so many women, after the Nazi victory in Germany, could boast of Hitler's favours—without having achieved the final intimacy with the Fuehrer. His reluctance to reveal himself completely, his anxiety not to become ridiculous—he, the national hero of the Reich!—acted as strong inhibitions.

After Hitler's triumph over his native opponents there were young girls and married women in almost every Gau of Germany who were "commanded" by the Gauleiter to entertain the Fuehrer during his visits to the district. It was known that he liked to be surrounded by them. These local favourites were very numerous; in their own bailiwicks they always figured as special friends or mistresses of the Fuehrer. But their relationship was of an entirely different nature; he needed other forms of surrender than the physical one.

"I remember," Luis Trenker says, "a reception in the Reich Chancellery in March, 1934. The Fuehrer was in the conservatory. I sat alone with a glass of beer. Luise Ulrich, the actress, passed by and persuaded me to go with her for the Fuehrer was 'telling stories.' As the large room was rather crowded, we stopped at the back. He sat in the middle, talking. A whole bevy of young women and men in evening dress were crowded around him; the whole *Stimmung* seemed to be familiar and homely. But Hitler was talking about guns and tanks and bombs; he explained that Germany needed even heavier planes, even more destructive bombs. The women nodded happily. The men made appropriate noises of assent. At Hitler's feet there sat, half-lying, half-kneeling, the wife of the director of the Nuremberg Museum, a woman of Baltic origin. She showed nothing but surrender, adoration, and enthusiasm. It was almost uncanny: dazzling women in evening dress, their shoulders and arms bare—and *he* spoke only of war, of destruction. The women, in their colourful silk dresses, fidgeted uneasily in their seats; they were excited—and so was he. There was a sultry, unhealthy atmosphere in the room; as if both Hitler and the women found a vicarious sexual satisfaction in this morbid talk."

It is to this period—between the coming to power in 1933 and the outbreak of war—that Leni Riefenstahl belongs. Of all the women whose names were publicly coupled with Hitler's, who were reputed to have entered into an intimate relationship with him, she is the only one alive. From 1936 almost to the end of the Nazi regime Leni Riefenstahl was accepted as "the" mistress of Hitler. She herself has done everything to spread this reputation. And she owed her own speedy rise in the Nazi hierarchy almost exclusively to it.

Later, in 1945, she declared in an interview, published in the "Saturday Evening Post," that she had never been Hitler's mistress—"he only respected me as an artist"—and insisted that they always discussed only the films and the stage when she was alone with him in her Berlin apartment.

Is this true? Luis Trenker, who knew Leni well, has a good deal to say about her. " Dr. Ernst Hanfstaengel, who used to be the Press Chief of the Nazi Party, but had to flee later from Germany to escape being murdered, told me some amusing details about Leni Riefenstahl," Trenker's report begins. " One day we were driving together along the Hindenburgstrasse, past Leni's flat. Hanfstaengel told me that one day he had been invited, together with Hitler, for a cup of coffee to Leni's flat. They stayed until two o'clock in the morning. Leni performed one of her notorious nude dances and, as Hanfstaengel put it, ' kept on shaking her navel in front of my nose!' "

This took place in 1933, but according to Trenker, already in 1932 Leni Riefenstahl had travelled to Munich in order to present Hitler with a copy of " Mein Kampf," in which she had annotated all the passages she liked with red pencil—while those she disagreed with, she had marked with black. They talked for hours; *she* was enraptured, *he* was deeply impressed " by this extraordinary woman."

" At that time," Trenker continues, " she was employed by Universal Pictures. She was expected in Hamburg—a whole company was to sail for the Balearic Islands to shoot a film on location. Udet and Fanck, the producer, were getting nervous, afraid to miss the boat—and Leni Riefenstahl was still absent. For days now she had been missing, and no one knew where she was. At last there came a telephone message. Hitler's private plane had just landed; Leni had been the Fuehrer's guest in a place near Nuremberg. She walked into the hotel lobby with a tremendous bouquet of flowers; her eyes seemed to gaze into the distance, her whole being was transformed; she wanted everyone to know that she had just passed through a wonderful experience.

From that day onwards wherever she was, in her room, in her cabin, in the sleeping car, she always placed a large photograph of Hitler with a personal dedication on her bedside table. Not one of her birthdays passed without a large basket of flowers sent by *him* with an appropriate birthday telegram."

When Mussolini visited Munich in 1938, there was one of the usual mass receptions in the Braunes Haus. Hundreds of prominent architects, painters, sculptors, directors, actors, composers, musicians, writers, poets, scientists, doctors, sportsmen, magicians, dancers and opera-singers were invited and obediently appeared, answering the call of the Nazi Reich. In the huge rooms there was a festive crowd. "Suddenly," Trenker relates, having been himself present, "solemn, breathless silence. The Fuehrer appears with Mussolini. Hitler is cold, stiff, self-conscious, as if seized by some rigid cramp from his topboots to his lock of hair; the Italian friendly, jovial, almost embarrassed. The two men walk through the room; all around them men and women stand frozen with arms raised in the Nazi-Fascist salute. Suddenly Hitler stops, looks at one of the groups, says a few words to Mussolini, and makes a summoning gesture. Leni Riefenstahl, blushing slightly, steps forward. She makes a deep curtsey, and Hitler introduces her to his guest. She is so excited that she does not know what to do or say. Her somewhat close-set eyes shine; in such moments she almost appears to squint. She knows what it means that of all this fine crop of artists and scientists she is the only one to be presented. For Hitler does not consider it necessary to introduce any other member of the distinguished company to the Duce."

"Another evening, in the *Braunes Haus* or in the Haus of German Artists," Trenker continues. "It was just like a beehive. Goebbels was surrounded the whole evening by a bevy of giggling, shimmering, chattering actresses. Directors of the State Theatres, in full evening dress, were rushing about, trying to pick up patronage. I was drinking a glass of beer with the camera-man Sepp Allgeier. About midnight the Fuehrer's confidential aide, Martin Bormann, appeared and told us to go down to the beer cellar where the Fuehrer and Goebbels were sitting; he said there were too many women and too few men down there. I did not feel like going. Leni Riefenstahl had complained because I had not greeted her and I did not feel like having a lengthy argument with her. Finally, I agreed. It was a huge room downstairs with an adjoining chamber, the door of which stood open. A single glance showed us Hitler at the table, with Goebbels, and about twenty women. But the table next to the door in the large room was also occupied entirely by women . . . and behind them a large empty space. At the back of the room you could only see evening suits, bald heads and smoke. I sat down at the second ladies' table and began to talk to Leni Riefenstahl. We were served *Leberknödel* (liver dumplings) and soup —for Hitler's court believed in rustic food. After a few minutes all *decolletages* began to tremble, all eyes brightened. I knew that *he* was coming. He sat down on Leni's left side. I had just explained to her that a difficult climbing tour which she intended to make in the Dolomites was madness—she would certainly be killed. Hitler must have heard the last few words, for he told her: ' Frau Riefenstahl, you must not risk your life too rashly; I have reserved very great tasks for you. You are going to make the film, '*National Socialism*,' and after that a really gigantic picture on *Germany*. Many will be able to realise your great and irreplaceable significance only after your death . . .' "

Trenker maintains that the " social " recognition which Leni Riefenstahl enjoyed was achieved by the artistic chances opened to her. The little dancer who had once been happy to secure an engagement in Innsbruck, was permitted to direct the film of the 1936 Olympic Games. She became an important personality of the Third Reich. Whenever she spoke of the Fuehrer, she quivered with emotion. Her eyes shone, thinking of the first night of her next film. The Fuehrer would present her with a huge bouquet in his box, would sit next to her while all around women watched her in envy. For she was the popular Leni Riefenstahl, the first lady of the Third Reich—at least so she thought.

She made '*The Victory of Faith*' and '*The Triumph of Willpower*,' the two pictures about the giant Party Rallies; she directed the '*S.A. Marches*' and '*The Polish Campaign*.' The negative of the last film she destroyed quickly before the Americans entered Kitzbühel. Nothing was too expensive, nothing too good for her films; everything and everybody had to serve her 'art' for she had to produce the supreme achievement. It was all for the dearly-beloved Fuehrer and for his wonderful ideas. In the Reich Propaganda Ministry they hated and envied her; but even Goebbels bowed low over her hand when she received him, cool and collected. She enjoyed this tribute; she was said to enjoy the full confidence of the Fuehrer. He presented her with a Mercedes car, and had a villa built for her—with a film studio in the garden. A bombproof shelter to secure the survival of her 'immortal pictures' was also included.

While she was making the Olympic film, she usually appeared in a snow-white costume; but during the shooting of the Polish Campaign picture she adopted field-grey—like her Fuehrer. She wore a soldier's cap, and a Sam Browne belt with a holster. She appeared in Danzig, in Warsaw, in Cracow; she rushed to and fro between the Fuehrer's H.Q. and the troops. While she was shooting the Olympic picture, she had forty cameramen and assistants at her disposal. Every evening there were 'story conferences' which had to be attended by her whole staff, dressed in dinner jackets. Every morning she appeared with 'Heil Hitler' and it was her last greeting at night.

Before Leni Reifenstahl set out for Poland to immortalise the 'heroic deeds' of the Fuehrer, there was a short but violent battle around her. The German armies had their own propaganda companies and the generals thought that a woman was out of place in a campaigning *Reichswehr*. When these objections were put to her, Leni fainted; then she knelt in front of the Fuehrer, begging him to let her go—and he yielded. This time, too, as always, she had a whole group of handsome young men around her—mostly Tyroleans. They were her 'personal assistants.' One of them—or perhaps several—were said to 'assist' her in the bedroom as well as behind the camera.

When the Olympic Games film was finished, Leni took it to the United States. She visited Hollywood 'on a special mission.' She appeared everywhere in the name of the Fuehrer and used Hitler's prestige for her 'conquests.' But Hollywood was neither impressed nor amused by her. She returned like a beaten general. Yet she succeeded in convincing both Hitler and Goebbels that her failure and her difficulties had been entirely due to the Jews. Her disappointment must have left a lasting scar. When Germany declared war on the U.S.A., she jumped to her feet in great excitement and called out in joyous ecstasy: 'At last, at last it has happened!'

To-day, three years after the end of the war, Leni Riefenstahl denies persistently that she has 'known Hitler well.' When she was interviewed by Budd Schulberg for the '*Saturday Evening Post*,' she told the well-known author of '*What Makes Sammy Run*': " Everything has been exaggerated so terribly . . . just because he received me alone a few times . . ." She added that Hitler had demanded of her that she should join the Nazi Party. 'No, Adolf,' she had told him (at least so she stated *after* Hitler died), 'I cannot do that. I am an artist. And an artist is a free spirit, not a party member.' She also denied having ever read 'Mein Kampf.'

But the charming Austrian actress, Dolly Haas, describing her last meeting with Leni in Switzerland in 1933, said that Frau Riefenstahl had told her: 'I gave up my left-wing ideas completely after reading *Mein Kampf*.' Schulberg, who visited Leni at the end of 1945 in her house in Kitzbühel, asked her whether she had really been Hitler's mistress. 'Of course not,' she replied. 'I wasn't his type at all. I am too strong, too positive. He liked gentle, cow-like women, like Eva Braun.' She begged the American journalist to help her—she wanted to be removed from the Allied black list. He told her politely that it wasn't his department to deal with these matters. Then she asked him whether he could give her a jerry-can full of petrol. He said that was forbidden. She smiled at him. " There was something strange in that smile," Schulberg wrote. " It was intimate, ingratiating and very self-assured. This is how she must have smiled at Hitler . . ." Now she wanted at least two pints of petrol. " It's forbidden," said the journalist. Thereupon her face became hard, nervous lines appeared in her panchromatic make up, she dropped the mask and was full of self-pity.

The character of Leni Riefenstahl appears to be full of contradictions. That she was a careerist and an opportunist cannot be doubted. Now that the Nazis have been defeated, she calls herself half-Jewish. It might even be true, Trenker writes. "According to the publication of Hanisch, Hitler's boyhood friend, the future Fuehrer told him that Jews had a different smell to Ayrians. 'I could never sleep with a Jewess,' Hitler said, 'for they stink.' Now Leni," Trenker continues, "certainly could not be accused of such lack of personal hygiene. But she often behaved in a vulgar manner, gesticulated with knife and fork, talked with her mouth full, picked her teeth and was not exactly a highly spiritual being. Eva Braun hated her—for very understandable reasons."

Leni Riefenstahl had a strong, elastic body; she could swear like a sergeant major, and had a hearty appetite. Her vitality must have been a constant stimulant to Hitler, who was strongly inclined to lethargy. Whether she was really his mistress must remain an unanswered question. But she is certainly the only survivor of the women who have played an important part in the Fuehrer's life. The other three, Geli Raubal, Unity Mitford and Eva Braun, are all dead.

5

WHEN Unity Mitford died in June, 1948, newspapers all over the world revived the memory of her tragic life, her entanglement with Hitler's magnetic personality, and her blind infatuation for the Fuehrer.

Sir Oswald Mosley's sister-in-law was a typical Nordic beauty, tall, fair-haired, full-breasted. She was Hitler's ideal of an Aryan maiden. She came under the Fuehrer's spell when she was a young, impressionable girl. While she worshipped at Hitler's shrine in Munich and Berlin, her brother fought in Spain on Franco's side. There can be little doubt that she was in love with the Fuehrer. But all his close associates maintained that their relationship was purely platonic. Some say, however, that it was Hitler who insisted on this. Perhaps he was content with her blind admiration. In this respect one recalls the story of how, in Nuremberg, he kept on staring at a young girl until she burst into violent sobs. " You'll never forget this day!" he told her. This seemed to satisfy him; perhaps he did not want anything more—or if he did, it was something deeper and darker, and far more evil.

In Hitler's relationship to Unity Mitford there was something of his characteristic Freudian ' Hassliebe ' (hate-love) for Britain. Again and again he called her into his intimate circle, heaped small attentions upon her which were intended as a homage not only to her femininity but to her nationality as well. In her he imagined he was subjugating the whole British Empire! But just because he saw in her not only the young and beautiful girl but also the Englishwoman, their relationship had to remain platonic. He would never have dreamt of betraying his small intimate secret (of which Eva Braun tells us) to her—precisely because she was British.

Soon after the outbreak of the war, Unity Mitford attempted suicide in the *Englischer Garten* in Munich. Like so many women around Hitler, she had a strong inclination to hysteria. Her life was saved and Hitler sent her in a special train to Switzerland. Relations with Britain had entered a new phase and Hitler hoped to subjugate the British Isles in other ways. She stayed in Switzerland for a while and then returned to England where she lived in complete seclusion until her death a few months ago. Whatever her relationship with the Fuehrer was, there can be no doubt that her life was blighted, and that her untimely death was caused by the 'magnetic personality' of Hitler and his unscrupulous exploitation of her uncritical admiration.

6

AND so we come to the fourth and last woman in Hitler's life—Eva Braun. All his other sexual adventures, even his dark passion for Geli Raubal, had been fleeting and unimportant compared to his relationship with this girl. We know now that these two were married on April 29th, 1945, shortly before the Russians occupied Berlin and the Third Reich collapsed in rubble and flames. A few hours later both of them were dead. Did Hitler want to legalise by this ceremony a relationship that had become a marriage to all practical purposes much earlier? Or was it the intention of the great 'magician of propaganda' to offer the world, a few hours before his unglorious exit, a last sensation? There was another, far more important and decisive reason for this dramatic wedding of which we will speak later.

Who was this Eva Braun? Who was the woman who managed to bind to herself, for almost ten years, one of the strangest and most uncanny figures of world history? What sort of creature achieved such a strong sexual hold over this 'mother's pet' (as he called himself), this incarnation of psychic abnormality, that he could never break away from her? She was, according to first-hand evidence, neither a *femme fatale* nor a high-class prostitute. She was a typical Bavarian girl, of medium height, fair-haired, slightly-built, with light-coloured eyes, and inconspicuous. She possessed a good figure, at least according to German standards—slim legs, heavy hips, small breasts, a slender neck. Her face was just average. She nearly always wore her pretty, dark-blond, wavy hair in artificial, loose disorder.

"I still remember clearly," Luis Trenker relates, "the evening when I met her for the first time. It was on a January afternoon that Professor Joseph Thorak, the well-known sculptor, rang me up at the Zugspitze where I had gone for some skiing, and invited me to supper in his studio. When I arrived, Thorak received me with evident pleasure but had only some bread and beer to offer. I was very hungry and because I was dressed in my old, rather shabby skiing suit, I did not want to go into a restaurant. I scolded Thorak, who was smoking a cigarette and seemed to be lost in thought. Suddenly he rose and said: 'Come on, I'll take you to a place where you'll get some good food.'

"A few minutes later we arrived at a villa in the Wasserburgstrasse. We were taken down to a basement room. There was nothing to eat there – but there was champagne and three women who appeared to be very bored with their own company. They must have already drunk few glasses for they were giggling in a silly way when we appeared. I knew one of them – she was Frau Almas, an art-dealer; a middle-aged woman who had played a certain part in Munich art life since the Nazis came to power because she acted as buyer for Hitler. Her popular nickname was "the Knick-Knacks Countess". Beside her sat two younger girls – one fair-haired, who did not look very attractive to me, and a prettier, dark eyed, brown-haired creature who was taller and made a better impression on me. There were no introductions, we just greeted each other and began to drink. The girls whom I took for typists or secretaries, wanted to know where I came from; when they heard that I had something to do with films, they began to enquire about stars and directors, camera-secrets and similar things. It was the usual flapper curiosity. The fair-haired, slim girl who was addressed as Eva by the others, tried to draw me out about Leni Riefenstahl; she insisted that I told her all the scandalous gossip I knew. At first I was reserved, then I related a few episodes of Leni's life, some of which I had witnessed myself—but I always broke off at the most interesting point or just before the story became too spicy. But that was just the point when Eva became really interested. We went on drinking. I forgot my hunger and finally we started to dance. I liked Gretel, the prettier of the two, much better; but somehow it came about that I was dancing almost exclusively with the smaller, paler, more pliant girl. It seemed to me that she needed affection far more than her friend. After midnight we went up to the drawing-room. My " little typist " took off her shoes and clung closer to me while dancing. Now and then I kissed her softly on her temple—and permitted myself the small intimacies which girls more or less expect from a man on such " bacchantic " occasions.

'It was six o'clock in the morning when I went back with Thorak to his studio. As I didn't know who these " typists " were who seemed to be living in such an expensive place, I asked the sculptor. Thorak had turned up his collar and was smoking silently. " Well, you've made a fine mess of things," he murmured. " You are not fit to be taken into decent company." " What do you mean?" I asked, laughing. " It was quite nice with those two girls, especially with the small one . . ." " The small one!" grunted Thorak angrily, yet with amusement. " You'll soon discover who the small one was—you'll be shot, my friend—that's all I can say . . ." " Sure I'll be shot," I smiled. " Isn't our country a highly moral one? But why exactly am I to be executed?"

Thorak's reply was nothing if not outspoken: "Because you were daring enough to pinch the Fuehrer's mistress." I had to laugh again. "You don't mean to tell me that one of those three is Hitler's mistress?" Thorak grinned. "Yes, it's the small one—she used to work in Hoffmann's studio. Now you'll believe me when I say they'll put you up against the wall . . ." Until this point it had sounded like a joke. "Well, what's her name? Who is this important lady?" "Eva Braun," answered the sculptor, "and you'd better keep your mouth shut or you'll really be shot!" And now there was nothing jocular about his voice.'

Luis Trenker continues: 'The next time I met Eva Braun was in Kitzbühel. I had taken the funicular up to the Hahnenkamm and enjoyed the wonderful crisp snow and the brilliant sunshine. On my way down I was passed by a sleigh which contained two ladies. One of them was Eva Braun, the other a full-bosomed, older lady, Mrs. Morell, wife of Hitler's personal physician from Berlin. They slowed down; I walked for a while at the side of the sleigh and talked to them. Eva Braun seemed to be in a very good temper. Finally the sleigh halted and the young girl got out. We walked uphill for some distance. Mrs. Morell remained behind.

'We found a suitable spot and sat down to rest. It was close to the road and many people passed by. Eva seemed to be less gay than in Munich—but there was a contentment about her; under the influence of the winter sun she appeared to be uninhibited and at peace. I asked her whether she still remembered the pleasant evening in Munich and how she had danced in her stockinged feet. She suddenly became stiff and icy; she looked at me and said: "Dear Herr Trenker, don't ever make such a remark again—don't ever refer to this. You hear me?

Never! Never again!" I must have looked puzzled as I hadn't expected such an outburst. " But it was all so harmless," I remarked. " Perhaps," she said. Now I noticed in her face, which suddenly looked haggard and pale, two deep lines which started at the corner of her lips. " You know nothing," she continued, turning away from me. " You don't know what a terrible tyrant *he* can be!" I was silent, and had the impression that I was sitting beside a deeply unhappy creature.

'Some passing skiers began to take snapshots of us. When Eva Braun noticed this, she became very frightened and hid behind me. " For God's sake, but this is impossible, let's go at once!" she cried, in sudden excitement. " But why are you so worried?" I asked. " It's an innocent pastime—all they are photographing is a young couple in the mountains. No one knows us . . ." " No, no!" she protested violently and rose. " Just think . . . if they should publish such a picture and it should come by some accident into *his* hands! I shudder to think of the consequences!" I understood her now and realised that she was right. I had no intention of getting into trouble with the almighty Fuehrer. But how dangerous a game I had unwittingly played, I discovered only when I read Eva Braun's diary and found out not only that she was under constant supervision, but also what radical methods Hitler had used to liquidate any possible rivals.

'We chose the easiest slope and I accompanied her part of the way. She was not a good skier and complained about her weak bindings. Before we reached the valley, she said goodbye and added that it would be better if we were not seen together in the village . . .'

There was, however, a different Eva Braun—who played the part of hostess at the receptions on the Obersalzberg, where she greeted guests and kept the conversation going until *he* appeared, and was the Pompadour in Hitler's court. "I remember,' Trenker tells us, 'a tea-party in Hitler's Berghof in Berchtesgaden. Eva Braun was the mistress of the house. She acted the *grande dame*, greeted everyone with condescension, directed the servants and the aides-de-camp and ruled the roost with her arrogant and precocious superiority. But immediately Hitler appeared, she became like a little mouse. She tried to make herself as small and as insignificant as possible. By some peculiar magic she seemed to shrink and disappear. She was still present, but no one noticed her. And this was a voluntary effacement, a retreat of fear—and she retreated not to the second but to the twentieth place of importance in the company . . .'

7

SHE came from a humble enough home. Her father was a teacher at an industrial college in Bavaria, an expert in woodcraft, a simple, lower middle class man. When, in the summer of 1945, a reporter of the '*Stars and Stripes*' visited the Brauns in Ruhpolding where they lived on a modest pension, hoping to discover some details about Eva Braun, her father appeared stiff and narrow-minded. 'When I heard the rumours,' he said, 'that there was something between Eva and Adolf Hitler, I wrote a letter to the Fuehrer and asked him what his intentions were and whether he intended to marry our Eva.' It's hardly surprising that this letter remained unanswered. But the naivety and utter lack of sophistication in "Papa" Braun is certain amazing. He and his wife had really no idea of the true character of their daughter; though Eva's mother was more of a realist, more ambitious and probably better informed. When Eva, working at the Berchtesgaden branch of Heinrich Hoffmann's photographic studio, wore leather shorts in her walks through the village, the strict father was most upset by such " immorality." But Frau Braun said: " Don't interfere; they suit her well, they show off her good figure."

According to lower middle class conceptions, Eva's sisters have risen in the world—they married well. One of them, Margarete, married Fegelein, who was a riding instructor and later an *Obergruppenführer* of the SS; Fegelein remained Hitler's favourite bodyguard until two days before the dictator's death. The other sister married twice—first a Jewish doctor who left Germany in 1933 and emigrated to the United States; then, as her second husband, she chose a diplomat, a section chief in the German Foreign Office.

Eva's origin, the small flat in Munich, the commercial college she attended, her work as a minor employee in Hoffmann's studio, explain her lack of social and artistic instincts. But she certainly showed little hesitation or uncertainty. After all, she had qualities of a different kind; she was probably the only woman in the world who could really satisfy Hitler sexually. In this respect she was powerful and secure, and she sensed her security. Whenever she was " showing off," she behaved rather vulgarly and unpleasantly . . . almost stupidly.

In February, 1939, she visited the Kitzbühel Grand Hotel for the first time. State Secretary Esser had reserved her a room, the best in the hotel. But Eva showed her disappointment at the " shabby old place " openly. Now the Grand Hotel was a really first-class house of European reputation, furnished tastefully, and containing a famous collection of old Tyrolean wood carvings which were scattered about the various reception rooms. Eva asked the proprietor whether the bedrooms were also furnished "in such an old-fashioned manner." He answered her with cool reserve, for he had no idea who she was. He merely mentioned that the King of Belgium, Princess Juliana of Holland and the Duke of Windsor had all visited his hotel frequently and were highly pleased with it. It was characteristic of Eva Braun that this information made very little impression on her; she was a snob in a different sense.

When the proprietor discovered that the Fuehrer, who had gone to Hamburg in order to launch the *Kraft-durch-Freude* ship "Wilhelm Gustloff," had telephoned "in person," he became much friendlier. The hotel staff listened in secretly to the telephone conversations between Eva Braun and Hitler. Thus the proprietor discovered that these two were on intimate terms, called each other "thou" and that Eva told the Fuehrer she found the hotel "not very nice." Thereupon the hotelier, who was a good judge of character, decided to find some special attraction for her. He sent for some skiing instructors and said to them: "Boys, keep her dancing—and give her a good time!" After that Eva Braun was surrounded every evening by handsome, tall, bronzed young men; they "kept her dancing," which made her far more lenient towards the Grand Hotel. According to her diary, whenever she danced, it wasn't the partner, but the rhythm, the movement which she thought important and in which she found pleasure. But in reality she was a very feminine woman, only interested in sex—she only felt well and comfortable if she lived in sexual tension.

A well-known German actor described a meeting with Eva Braun in 1943, in war-time Berlin. "The mood of the people in the German capital became gloomier and gloomier every day, which was easy to understand. The evenings were tense and heavy, the large-scale raids had begun to hit Berlin with their full power. In order to escape this depressing atmosphere, at least for a few hours, I often went to the theatre while I was rehearsing a new play myself. One evening I attended a deeply moving play which dealt with the tragedy of a mother who had to hide the existence of her illegitimate child from her narrow-minded husband. The very difficult part was brilliantly acted by Käthe Gold, one of the finest German actresses, and the play itself was really a first-class piece of theatre, not to mention its real literary merit. I was much impressed by both play and performance. At the end of the performance I met Eva Braun and Mrs. Morell in the corridor. Eva looked very youthful and was dressed simply. 'Well, what do you think of it?' I asked her. She replied quite coolly, without any display of emotion: 'It was tripe!' I was so perplexed that I could only stammer: 'But Käthe Gold! Käthe Gold!' Eva Braun threw her head back and said: 'I don't like her, I can't understand what people see in her. I didn't like her performance at all . . .' She gave me her hand to kiss and walked on . . ."

This description probably gives a good insight into Eva Braun's mental and spiritual makc-up. For curiosity's sake, we might add here the picture which Frau Erna Hoffmann, wife of Eva's former employer, painted of her. She had known Eva well, for Hitler's favourite was a frequent and welcome visitor in her house. We know from the diary that it was Heinrich Hoffmann who acted as Hitler's procurer in Eva's case; and we also know that she had once made Erna Hoffmann extremely jealous, by " snapping up " the young Viennese painter who had only been invited to the Hoffmann house because

Frau Erna liked him more than she should have done —for he was no Nazi.

When a French reporter called on Erna Hoffmann in the Tyrolean village where she lived in poverty (for her lover, a Greek singer, had left her and taken all her valuables with him), she told him that not until the marriage certificate had been published was she able to believe that the Fuehrer had really married Eva Braun. Eva Braun she said, was by no means Hitler's most influential feminine adviser, as the newspapers tried to make her out after her death. This is probably true. Eva Braun was Hitler's favourite audience; but she never offered either advice or suggestions.

" Eva Braun," declared Frau Hoffmann, " was one of Hitler's occasional mistresses. Their relationship had no real significance. Eva was a pretty girl but somewhat hysterical. When Hitler sent for her for the first time, she thought her great hour had come. She saw herself as the Madame Pompadour of the Third Reich. But Hitler dropped her without any explanation. Eva often came to me, and cried and sobbed because she was so unhappy and wanted to commit suicide."

" Did Eva Braun love Hitler?" the reporter asked. " Loved? That would be saying too much," Erna Hoffmann replied. " She was more impressed by the Fuehrer than by the man. But she usually told me stories about Hitler, the man. He always received her with the same ceremony as all his other mistresses. Everything was prepared for the rendezvous: a loose ' Valkyrie night-gown' an open log-fire and champagne. During these meetings Hitler never uttered a single word. After he had withdrawn, he always sent his aide-de-camp to Eva, telling her he would fulfil one of her wishes. The first time Eva replied that a cup of coffee would be very nice. 'I was a silly goose then,' Eva told me. Later she asked for jewellery. Now and then Hitler took her on one of his tours. Once or twice he presented her at an official reception. When he began to drop her, she started to write him letters which he never answered. Then she found consolation elsewhere. But she did not want to give up her hope of becoming Germany's Messalina. By 1940 it seemed that Hitler had finally broken with her—though Eva bore him a daughter. It was not quite certain, though, whether he was the father. I know nothing of a supposed son . . ."

Now, in full possession of the evidence, we can only say that the above description represents the testimony of an ignorant woman with a primitive imagination, tinged by the jealousy she felt against her successful rival. In any case, Erna Hoffmann's words are characteristic of the behaviour and attitude of the leading Nazi circles. Even Hoffmann's wife was ill-informed about the real relationship between Hitler and Eva Braun.

All the available facts disprove her statements. From 1938 onwards Eva Braun appeared constantly in Hitler's company. Josef Thorak, the sculpter, related that Hitler even sent for her whenever there were important discussions in Berlin; he caressed her in the presence of his most intimate followers. She had to be available always and everywhere; and when they were apart, he telephoned her every night. The most important men in Hitler's innermost circle—men like Speer, Bormann, Hoffmann—spoke of her only as Hitler's mistress; even Himmler treated her with the utmost respect. And finally—Hitler married her.

In spite of their evident mistakes, Erna Hoffmann's romantic and sentimental stories are interesting—for they prove Eva Braun's almost superhuman discretion, her absolute silence which nothing could shake. This is the only explanation of how Hitler's relationship with her remained secret to the very end, not only from the general public, but even from the relatively initiated people who were very close to him. Most of those in his wider circle of collaborators knew nothing at all about Eva Braun, although she was so often in his company and had her own suite of rooms.

8.

WE HAVE now seen something of Eva Braun as the world saw her. But what was she *really* like? was there another, different person whom only Hitler knew — or perhaps the unhappy young Viennese painter who appears as almost the only decent character in her diary? The man who paid so high a price for his light-hearted and impulsive infatuation for this Bavarian girl . . . The diary itself answers these questions—with a frankness, a shamelessness, a lack of inhibition which are almost breath-taking. Some passages might even be called obscene—except that the total lack of emotion, and the absolute absence of shame gives this diary an abstract and unique quality.

Of Hitler's sex life, of Eva Braun and their relationship, the diary tells us practically all that can be told. But at the same time it draws a terrifying picture of the leaders of the Third Reich—always from the feminine angle, always from an erotic point of view — which fills us with loathing and nausea towards these perverts.

Her talent for discovering and describing these men and women who were morally insane is the more surprising for it appears to be fully proved that Eva Braun knew no man intimately before she met Hitler. She was a virgin—even if she was not so completely innocent as some might think when she went to her first rendezvous with the Fuehrer. The vulgar and outspoken instructions which her boss, Heinrich Hoffmann, gave to her, prove that. And she put on her very best underclothes—certainly not the action of a girl keeping an innocent date; but neither was it the behaviour of an experienced woman, hoping to become "the Pompadour of Germany." It was purely the reaction of a very feminine woman born for physical love, immediately excited upon entering an erotic field or even touching its periphery. Eva Braun senses all the possibilities of sex with an absolutely sure instinct. She is not the partner who enjoys all this; she is the great giver, the ideal mistress who satisfies completely, the perfect partner in any erotic game. She does not fail even under the most difficult conditions, when she comes up against an apparently impassable barrier of obstacles, of inhibitions, and of physical and psychological reservations.

Her difficulties began with the simplest things. She told us how Hitler hated to be seen without his clothes; a characteristic which all his biographers and friends agree upon. But he had nothing against *her* being naked. Eva relates how he sat there, completely dressed, the neatest man she had ever seen, watching her attentively as if he wanted to remember every movement. One can feel the sultry, unhealthy atmosphere of the whole situation. Again and again he wanted to watch her in the nude. He asked her whether she did not feel warm dressed—and she understood and undressed every time he gave her a hint. He also evolved some physical exercises for her. These must have been, we are bound to feel, obscene in some way or other—for everything in this atmosphere was unnatural, perverted and repulsive. He also liked to watch her while she took a bath—and Eva felt that, " happily " he could never have enough of such a sight. Watching her during her exercises, he told her that she was " very elastic," always the same words; he called her attention " to various details "—and one acquires the impression that he knew her body better than Eva herself.

Occasionally there is a touch of the ludicrous and grotesque. Eva describes a " charming evening " on the Obersalzberg. They had their evening meal; after which Hitler, completely dressed as always, took a foot-bath, sitting on the edge of the bath-tub. The water was mixed with all kinds of bath-salts to imitate seawater as closely as possible! While he relaxed in this way, Eva read to him from an old book about Alexander the Great . . . Could anybody invent a more devastating caricature of a dictator? This was the " great magician "—at home. He had over-sensitive, ever damp, ever painful, blistered feet for which there had to be a dozen perfectly polished pairs of shoes ready every day. When he rose, he " inspected " these shoes, walking up and down in front of them and looking at them as if they were living creatures. Here, of course, the monumental *petit bourgeois* betrayed himself; the man with whom it was " impossible to discuss his parents." Eva Braun realised that he had complexes; that he would really have liked to have been a *homunculus*, coming from the mysterious void and returning into nothingness . . .

The " visual hunger " and " the almost pathological need for amusement " of which Konrad Heiden speaks were later fulfilled almost entirely by closely watching his mistress. It seemed as if he wanted to explore her body incessantly; one might be tempted to conclude that it was the only female body available to him for such a purpose. He told Eva that she was not the most beautiful woman he knew but she certainly possessed the "secret of eternal stimulation." Was this just bombast—or the deep-rooted cry of a man who had always been cheated of the final favour? Eva Braun tells us that he was "shy." He always turned away when he undressed. And yet she never looked at him, not at his body. The hypnotic power, for her, was hidden in his eyes.

Adolf Hitler did not like to waste time in sleep; all his biographers agree that he was "full of a sharp world-hunger." To be diverted, to look and watch and see, was his deepest necessity. He hated sport and exercise. His moderation in food and drink was determined by experiences of which we must speak later. His weak lungs made smoking impossible. But at night, in the company of a woman who knew his strange tastes, he came to life. Konrad Heiden has stated: "We must always remember that his dignity was something assumed and that licentiousness was his true nature."

What was this man really like? For we must realise that no real, truthful pictures existed of him. "No photograph contains this dual personality, trembling for ever between two polarities . . . We only have photographs of certain conditions in the raw material called Hitler." Professor Max von Gruber of Munich University who used to be the leading anthropologist of Germany, described Hitler to the State Prosecutor in 1923 in the following manner: "Face and head show an inferior race, of mixed origin. Low forehead, unprepossessing nose, broad cheek-bones, small eyes, dark hair; expression of face shows a man not in complete control, but someone who is madly excited. Finally, an impression of blissful self-assertion." Other observers have noticed the more subtle marks of hysteria in Hitler. Someone who knew him well described him thus: "The unnatural change from rigid self-control to quivering fury; the studied attitude of an emperor at the front and the somnolent huddle at various festivals; the convulsive walk on parade and the clumsy, loping gait in the garden—one would expect him to carry an umbrella — all show this duality. Always the sudden, quivering transition from one to the other of these sharply opposed poses, the confused jump from the true to the artificial nature. He can cry when he wishes and laugh when he so desires, can pretend to be moved and manufacture outbursts of fury; can fire himself artificially or turn rigid. Though he produces tears on every occasion, he brags quite jovially that he hasn't wept for years. While outwardly he may show perfect calmness, he is preparing an outburst of maniacal fury within, rising until he ejects it from himself."

Has no one seen this man without any of his numerous masks? If there was a single human being, it must have been Eva Braun.

Eva had a strange, almost photographic gift for describing experiences and human beings alike; fundamentally she paints a more ruthless picture of Hitler than any of his enemies. She never passes judgments about anyone and if she now and then makes a remark (as about Goebbels), it is in general terms. But such opinions are seldom expressed. She describes everything visually; much of what she writes reads like a film-scenario, a series of sound-pictures. But it is exactly these sequences that tell us more than a thousand judgments. She leaves it to us to draw our own conclusions.

Eva Braun, without doubt, clearly realised Hitler's

real nature. But she seldom expressed this knowledge and then always in the form of hints. She does not give us an answer to the problem of his " face." Yet she offers exhaustive information about his " being "—and this in the manner which always leaves the question open as to whether she loves him or not. There are many indications that she was afraid of him. On the other hand the desire to be " Germany's secret queen " was certainly very strong in her. She was never consulted in political matters, nor did she have any ambitions in this field. She was astonishingly indifferent to the Fuehrer's great decisions. Her main interest, indeed, was guarding herself against other women and possible rivals — women like Leni Riefenstahl or Magda Goebbels.

In the hotel lounge at Kitzbühel she put up a small, post-card size picture of the Fuehrer near her usual table—so that she could always see him. But in contrast to Leni Riefenstahl she never spoke of him with " ecstatic eyes " or with the adoration a young girl usually has for her lover. Nor could one notice any sign of embarrassment when she spoke of him. She sounded more as if she were discussing some expensive, powerful and mysterious machine—or some building which the average person could only see from the outside, while she (and this gave her a feeling of superiority and security) could enter at any time. Now and then she was guilty of small trespasses against the commandments of her " divinity." She asked an officer, for instance, to bring her some fashionable shoes from Paris; but the Fuehrer was not to know about it as he did not want her to wear such shoes. Hitler hated red nail polish; but when Eva was away from him, she used it happily enough. She only deceived him once in her life—when she came near to falling in love: the magic

Nazi circle was broken by a young man, uninterested in politics, and she lost her head.

The young man was Viennese. His name was Kurt and he had no idea of Eva Braun's identity. He abused the Nazis and asked whether she was one. Into the middle of their budding romance falls, like a bomb, a letter from Hitler which makes her " quite mad." Kurt followed her to Berlin. What happened then is best read in Eva Braun's diary.

But the excuse she uses is significant: " Well, after all, I wanted a *man* for once!" This remark and a few other references place Hitler's unofficial marriage to Eva Braun in a strangely clear light.

It was a long time before Hitler's mistress reached this stage, which in itself proves that she had no experiences with other men. Only in 1940 did she discover " the secret " and then it was a book that called her attention to it. She was reading Stefan Zweig's " *Marie Antoinette,*" — a strange book for her to read and one that no good Nazi could have opened, for Zweig was both a Jew and a courageous opponent of Hitler. She was especially interested in the chapter about Louis XVI's slight physical defect which made his marriage so difficult. Eva was a healthy Bavarian girl, with well developed sexual instincts. She soon sensed that something was " wrong " with the Fuehrer; but while Marie Antoinette had a very experienced mother to advise her, Eva had no one to whom she could turn. She therefore thought of borrowing a work on anatomy, and wondered if she could ask Dr. Morell for one— but she hesitated before taking such an indiscreet step. After all she did not even know whether Hitler realised his own defect, whether he knew that he was " not normal." Eva also speaks of the three women whom Hitler had loved and expressed her conviction that none of these had ever told him if they had noticed it at all. She admits, too, that Louis XVI's " troubles " were far more serious than Hitler's. For Louis XVI and Marie Antoinette wanted children while she had no such intention.

Now here we arrive at perhaps the most delicate and certainly the most interesting problem in Hitler's sexual life. Yet it was, in many ways, a ridiculous predicament—a kind of poetic justice ironically devised by Mother Nature.

For those who have not read Stefan Zweig's brilliant " *Marie Antoinette,*" it should be explained that Louis XVI suffered from a serious form of phimosis or paraphimosis. The former means that the opening of the foreskin of the genital organ is too narrow, sometimes so much so that not only does it refuse to retract, but becomes so rigid that micturition is hampered. In paraphimosis, the foreskin is not so abnormally narrow, and retraction is possible, but it may suddenly refuse to come back into place; it locks itself at the base of the glans and resists all attempts to bring it back to cover the latter.

Louis XVI's difficulty was finally normalised by a small operation which was hardly dangerous even in the eighteenth century. Hitler suffered from the same trouble. But the operation which would have made him completely normal, and would have removed the bitter and persistent inhibitions which interfered with his relations with women, was basically nothing but—circumcision. And of all the people in the world the Fuehrer of the German Reich could not possibly permit this operation to be performed on himself. He, who hated and tortured Israel, who was the scourge of the Jews, could not submit to the same surgical intervention which every faithful Jewish child undergoes at a tender age! Adolf

Hitler, whatever his troubles and difficulties, could not be circumcised!

This predicament explains his relationship to Geli Raubal, and makes it perfectly clear why he was able to promise his step-sister that he would not " seduce " her; why he used the expression " demi-vierge " and why the young girl whom he approached with certain demands, had simply called him a " horrid fellow "; why he only frightened her when she realised his dangerous nature. It also shows why another woman had declared that he had proved himself " little of a man." The theory of his impotence, however, must certainly be abandoned. He was not impotent; Eva Braun became pregnant and bore him a child. She describes him as shy and difficult but ascribes this to certain bad experiences.

Did his whole twisted character owe its development to this single fact? Was this the reason why the most natural things became a problem for him? Eva refers again and again to his shyness and his inhibitions.

We are justified in placing the relationship of Hitler and Eva Braun on the lowest and most primitive human level. There was no " mating of souls " involved. Even loyalty and faithfulness were lacking on both sides. It was a sensual " community of interests " in which Hitler played the more pleasant part while Eva Braun obtained the greater profit.

True, he very often spoke to her about his political plans, his intentions, his attitude to France, to Petain, to Mussolini, Ciano, Roosevelt, Churchill and the Pope. He explained to her what he planned to do with Poland or with France. But this did not mean that there was any intellectual contact between them. These were not conversations but lectures—which he delivered to any and every person who entered his circle. This pathological eloquence, this flood of words, this constant self-explanation must have originated in his youth when " clumsy and shy," he had been tortured by the constant reminder of his lack of success in personal relations. The man who " wolfed down half-a-dozen cream buns in the Carlton Tea Rooms in Munich, who watched the juggler Rastelli happily in the Deutsches Theater, who went twice a day to the cinema and put his arms around the lady whom he took with him "—this man became a national hero, petrified into complete rigidity; a man who concentrated all his wild greed and lust of life upon one object—Eva Braun. He had another mistress—Germany. And he probably possessed more power than any one man had ever possessed before him; he enjoyed it, too, with a demonic intensity which was not entirely free from sexual elements. He got his beloved war, too; even though it destroyed him in the end. By 1944 he had become gross and shapeless; Eva Braun discovered that he had reached the end of his tether and was unable to make love. All had been exhausted, everything was nearing its end.

About this time even Eva Braun, who had never shown the slightest concern about events in Germany or the world at large, began to show some signs of alarm. We can sense from the diary how she began to be uneasy, how she suddenly noticed that whole cities were turned into ruins, and death stalked almost unchecked in the Reich. She consulted Hitler's personal physician about his physical and mental decline. At last even this callous little woman, who lived only for sex, was caught by the spectre of the German tragedy. A man died in a Berlin shelter and the doctor stated it was " of fear." But even at this stage, Eva still insisted that a high civil servant who told her an anti-Nazi joke " must be executed." She was a true Nazi, and this was one of the reasons why she remained with Hitler to the bitter end. She was the only woman permitted to enter the " Speer Circle " in the Fuehrer's H.Q.—the circle which, according to General Jodl's testimony at Nuremberg, was a cross between a concentration camp and a monastery. She stuck to the Fuehrer with the loyalty of a camp follower. And she remained with him because it appears that she bore him a child which was to be made legitimate before they both died.

Konrad Heiden, in his biography, maintains that Hitler had children. This is more than possible. In Eva Braun's diary there is a long passage reproducing Hitler's ideas on this point. After careful investigation of the diary and other sources, many experts are inclined to believe that Eva Braun bore Hitler a son in Dresden in 1942; that Adolf Hitler married her shortly before his death in order to give the illegitimate child a name and to bequeath him a heritage of world politics.

One might object that all this was hardly " normal." But was Hitler normal? Eva Braun shows us that he sometimes had serious doubts about his own sanity. Again and again she found him reading the " Textbook of Psychiatry," by the Swiss Professor Bleuer. Eva herself was afraid of him. She once told Professor Thorak that she was never sure whether he would murder her one day. No one was safe from him. She was probably thinking of Geli Raubal's fate. But if anything like this were to happen, she added, she would be certain that Hitler, though the murderer, would not be responsible for his actions. For this reason, too, she tried to avoid making any enemies. In her diary she repeatedly writes that she sometimes doubts Hitler's sanity. She quotes many instances of his strange behaviour, and reasons for her belief in his unbalanced mind.

Of Hitler's " haunting hysteria " we have already

heard from Rauschning. In the years 1934-1938, this most intelligent observer painted a horrible picture of the dictator: " Fear overtakes him. The happy emotions which have filled him a moment ago, disappear. He is persecuted, he is spied upon. He hears whispers which fall silent when he approaches closer. People look at him more and more rarely. They are talking of him. What are they saying? They cannot be jokes any longer. They all look gloomy. These people have some evil intention." This description of persecution mania which psychiatrists ascribe to paranoia, cannot be surpassed. " In his delirium," Hermann Rauschning continues in a different passage, " Hitler, in the rare, private hours spent in his mountain eyrie, feels himself to be the greatest genius of his people and the greatest future law-giver of mankind. . . . Is there a single field of human action in which he has not produced revolutionary ideas? He is greater than Frederick the Great, than Napoleon, than Caesar!"

We could go on quoting for several pages; we could produce a great many witnesses, besides Eva Braun who perhaps knew him best. It is, therefore, the more surprising that the Nazi Generals who were tried in Nuremberg declared that they had never doubted the Fuehrer's sanity. That a Baldur von Schirach or a Streicher followed him to the bitter end, can be understood. Hitler's satellites tried to outdo him in "surpassing the depravity of the renaissance." But the German generals were of different origin and education—and yet they blindly refused to accept the fact that they were led by a madman.

What was the origin of this madness? There is in " Mein Kampf " a chapter that has been comparatively neglected in analysing his life and character. It begins with a discussion on syphilis, in which

Hitler demands the sterilisation of incurables. Konrad Heiden noticed the " over-sensitiveness, one-sidedness and irritability " with which Hitler " touched on wounds and castigated vices." He called prostitution " the disgrace of humanity." Its best remedy was, according to him, an early marriage for all men. But Hitler himself did not marry. He maintained that "severe training," much physical exercise and sport was a help, that " a youth hardened in such a way is less subject to the necessity of sensual satisfaction than a bookworm whose head is crammed with purely intellectual food." But he himself never took any exercise. And Eva Braun's diary shows us that Hitler acquired these views only after it was—too late. When he writes about venereal disease, he loses all sense of proportion and indulges in rhetorical outbursts which are unusual even for him. Some observers feel that certain personal experiences must lie behind these feverish fantasies.

Yet in " Mein Kampf " Hitler fights not only syphilis but also prostitution. He describes the temptation, the foolish fall of a drunken man or a careless youth with such strong feeling that even a superficial reader must sense the personal experience behind it. Now Eva Braun's diary gives us a clear answer as to how this violent, almost mad fury against prostitutes was planted in Hitler's mind. And Eva herself, though she tried to reassure herself, was never certain whether Hitler had been healed of the infection which he acquired on his seventeenth birthday—or whether his outbursts of madness, his unbalanced ravings were due to the " pale spyrochetæ . . ."

The diary, however, gives us no answer to the question of whether she loved him or not. She often mentions, rather pleased with herself, his " love " for her. But these declarations appear to be somewhat conventional. She was moved when this " Spartan man " gave her detailed advice on cosmetics, or when he asked her to wear chamois leather underclothes . . . But love?

A school-friend of Eva Braun, who was interviewed by an American journalist, related that Hitler's future mistress had been " shy, reserved, and sensuous " as a child, unable to make real friends, and therefore not very popular at school. In the diary she appears almost incredibly cold as a human being; nothing touched her soul and her uneasiness during 1943 and 1944 was caused by events in Germany which started to make her nervous. Hearing of concentration camps, she was " interested "; when war broke out, she felt frightened, but soon recovered her poise; a report about thousands of executions which came into her hands by accident, left her completely cold. Even when the only man who managed to break through her reserve died, she merely wrote a few callous sentences in her diary.

But she felt linked to Hitler by a magical, inescapable bond. After her short love-affair with Kurt, the only experience which ever approached a real love-story, she writes " there are no other men for me." When she exclaims: " I was so happy in those days . . ." one is tempted to believe that it was the happiness of a normal young woman. Hitler gave her a ring with the inscription: " Until death do us part!" Yet one is dogged by the impression that there was nothing natural, normal or clean in their relationship. It had a sultry depravity and insincerity—like two moles huddling together in a dark hole. The fact that she made such detailed notes of everything she experienced with Hitler was in itself a kind of depravity.

The question must be asked: Was Eva Braun truthful in her diary? Perhaps she gave free reign to her lurid imagination and invented all these tales? Perhaps she just made them up from her own dreams and not from reality? As far as the sexual relationship between them is concerned, it is the opinion of all those who knew her that she was truthful. Her diary is crowded with scenes and pictures in an unbroken sequence, which complete each other and correspond to the opinions of those who knew Hitler well, so that it is difficult to doubt the truth of her descriptions. They carry the stamp of actual facts and experiences. Some people may be shocked by this insight into an uninhibited underworld. What Eva Braun tells us is the more terrible because it concerns not only the man who brought one of the greatest disasters to the modern world but also his whole depraved and morally rotten circle.

She described events coldly and without compassion; she saw everything with unblinking, icy eyes. She possessed a one-sided intelligence, and an abnormal lack of emotion. She was a mixture of lower middle class conventions and uninhibited lasciviousness; she was both primitive and crafty; she often appeared stupid but she also possessed incredible shrewdness. Beyond doubt she had a considerable though utterly untrained literary talent. Her presentation was often amazingly intensive, especially when she participated in events which aroused her sensuality, or in her notes on her adventures with Kurt, which show a refined and calculated subtlety.

She also possessed an extraordinary memory. She reproduces Hitler's long soliloquies about a variety of questions which did not interest her at all, with remarkable fidelity. We recognise Hitler's style and must, on the whole, accept Eva's diary notes of these endless speeches as faithful and accurate. Here and there we find inaccuracies but these do not matter in evaluating the diary; for this is not a political document but a contribution to the moral history of the Third Reich. It is certainly no accident that the diary does not contain a single real political "sensation." Hundreds of small references and hints add to the already known political tenets of the dictator. But even these parts contain interesting details. It is also advisable to keep in mind what Eva Braun considered worth noting down apart from her personal experiences, such as her visits to Hess, dinners in Goering's house, discussions at the Hoffmann parties, the very private gatherings around Hitler or the breakfast with Goebbels. All these encounters are recorded because Eva Braun was curious, and because those she met had certain reputations. But there are also long passages from Hitler's rather abstract speeches and lectures, a "big do" in Funk's house, the visit of Hitler's court astrologer and several similar details in which we cannot trace her real interest.

She behaved modestly enough in the public. She wanted to draw no attention to her own person and very seldom did. But she still felt herself to be the "Gretchen" of the Third Reich. True, her "Faust" was possessed by vastly different demons from Goethe's, who triumphed over Mephistopheles.

PART TWO

EVA BRAUN'S DIARY

EVA BRAUN'S DIARY begins in the last months of 1937. At this time she was already very intimate with Adolf Hitler, as the very first lines prove. 1937 was really the last year of untroubled peace, for 1938 from February onwards stood under the steadily growing threat of war. In 1937 Hitler was preparing the Austrian Anschluss. Inside the country the struggle for power between the army and the party reached a decisive stage. Goering and Himmler had succeeded in removing von Blomberg from the post of Minister of War. General von Blomberg had married, in somewhat mysterious circumstances, a Berlin prostitute who had served seven prison sentences for moral offences; this led to the serious weakening of his position. General von Fritsch, Chief of the General Staff, was involved by the Gestapo in a scandal which cost him his job. He had been accused of homosexuality but was able to prove his complete innocence. In spite of this the Blomberg-Fritsch crisis marked the beginning of the capitulation of the Wehrmacht to the superiority of the Nazi Party and the SS; the German generals were in no way able to fend off the shameless Nazi attack on von Fritsch.

Eva Braun knows nothing of all this. While Hitler attends the "Party Day of Labour" and demonstrates the power of the Third Reich in mass parades, Eva indulges herself in reminiscences. This must have been roughly the anniversary of their first meeting. Already, though the world was far from recognising the terrible danger of this man, Hitler was reading Professor Bleuer's "Textbook of Psychiatry" for weeks on end, and talking to his mistress about "genius and madness."

Eva Braun's diary was, of course, written in German and she used a good many Bavarian idioms and slang expressions. We have tried to reproduce these in English as far as possible, but of course no literal translation is possible.

No date, probably September, 1937

Wednesday night. The moon shines through the window. I cannot sleep. To-day I celebrated the anniversary of our acquaintance, of our first caresses. How much has happened since then, how my life has changed! "He" of course, did not remember it, but he has really more important things to think of. I cannot expect such things. But I see everything as clearly as if it had happened yesterday or to-day. When he sent me word through H. to call at the *Braunes Haus* at six o'clock, I thought at first he wanted to dictate something or there was some other work to do—I, poor wretch! Fortunately Heinrich understood the whole matter far better—or perhaps they had discussed it, though I don't believe *that*. In any case it was good that H. prepared me for it. He didn't do it very tactfully; more like a sort of joke. Like the man who is asked to break the bad news to the wife, and asks "Are you the widow Mayer?"—H. told me: "Take a bath and wash yourself all over. You know, the Fuehrer is mad about cleanliness; and he's taken a fancy to you." Of course, I was amazed, and then I can't really remember whether I was very proud or just afraid. Of course, I had seen "him" several times but always at some distance—and now he was supposed to have taken a fancy to me! I noticed how he looked at me when he came; I was wearing my close-fitting pullover, the blue woollen one . . . and, to be quite honest, I stretched myself a bit when he came in; what girl would not have done the same, wasn't it quite innocent? And then what? I did not put on the olive-green dress with the black bodice; it has become a sort of national relic. Nor my silk underwear, it was the only set I had at that time and the slip had become rather short and a bit too tight . . . I had not a drop of perfume at home and that was my good fortune for he hates perfume, at least he dislikes it. He always curses " women who smell." I took the tram; I was not supposed to go in by the main entrance. Down at the side-door an aide-de-camp in full uniform awaited me. I followed him, floor after floor, like in a film; then a steep staircase where it was quite dark so that he had to lead me by the hand. Then across an open gallery which gave a fine view of the whole town and then through a French window again into a dark corridor; and nowhere a human being, as if the house were deserted. I would lie if I said that I wasn't frightened. But I was also madly curious. After all, it was a crazy adventure. At last we entered a room full of curtains and from there we came into a normally lighted room, his ante-room, done up in white and gold. From there a large folding-door opened into *his* study. In front of the door there lay a wonderful Alsatian who later completely accepted me as his mistress; they told me he never behaved the same way to any other woman. I remember how I started when the door suddenly opened; but only the white-haired Schwarz, the Reich Treasurer, came out, loaded with files, and looked at me grimly through his pince-nez. Then *he* came, smiled at me kindly, took me gently by the hand and said: " Come on in. It's more comfortable in my room." He went ahead, the room was large and well-lit and pretty. I can still see it as if in a dream. How often have I seen it since, but never so full of brilliance as on that first great day . . . It was homely, with illustrated papers, books, flowers, a huge desk and a piano, and a fireplace in which the fire was burning—just like a fairy-tale. And then he said: " So you are Fräulein Hoffmann?" And I almost lost my voice and my heart raced as I said: " Heil Hitler, my Fuehrer, but I am not Fräulein Hoffmann, I am Fräulein Eva Braun." How often he has laughed about it that I said: " Heil Hitler, my Fuehrer," instead of " Heil, my Fuehrer," as it was prescribed! The first time, too, he laughed heartily and looked at me affectionately with his deep-blue eyes so that all my fears vanished. The dinner was wonderful for *he* is the most charming host you can imagine. Since then I have learned that food was usually very modest at his table but my first visit was a solemn occasion and there was a festive banquet. Hors d'œuvres with Russian caviare, " the best Little Father Stalin can send us," he said; some fresh salmon, a special delicacy; then woodcock which was so tender that it melted on your tongue, and wasn't a bit " gamey." The wine *I* drank alone, *he* only poured it out. First a wonderful ice-cold Chablis, then a Perrier-Jouet and finally an Haute Sauterne. Afterwards he made the coffee himself and I talked and talked for I felt the sweetest tipsiness I had ever felt in my life. I lay on the sofa in seventh heaven. I woke up in bed and now I am thinking and thinking for after all it was the first time and the memory clings to the end of your life. But I could not remember what happened and I still don't remember.

October, 1937

One-armed Herr Amann came to call on me and wanted to give me instructions " in the name of the Party " as to my relations with *him*. I have become practically a party institution! But I refused bluntly. He said I ought always to be careful what I said to the Fuehrer and should always discuss these matters in advance with him. This, of course, is nonsense. First of all *he* seldom asks me any questions; though a little while ago he enquired why people believed there was a controversy between Goering and Goebbels—but this happens very rarely. And then I always tell him the truth, the unadorned truth; and for that I need no Herr Amann. If the Fuehrer wants to discover something through me, then he's going to discover it. They won't get me to spout *their* views to him. Perhaps in the end they would like to make up a special newspaper for him as it used to be done for the old Emperor Francis Joseph. Fundamentally no one understands *him*. Not so long ago he told me how they always spoke of him as a political genius—" but I am not a *political* genius, I am just a genius, *the* genius of my age and this age will be called after me." He also said that he would have achieved world-wide success in any profession and in any position. " I am a politician simply because this age can only be re-cast through politics and perhaps through war, given its proper shape. If war should be unavoidable, which is a possibility, then I'll be the most important general of my age." And I was silent at these words because I felt so small; after all, what could I have said? Perhaps he was right, perhaps not.

The first Nazi *Putsch* in Austria, staged in July 1934, was always officially represented in the Third

Reich as an entirely interior movement of the small neighbouring country. According to official Berlin declarations Germany had no influence on its birth and development. These statements have been disproved long ago. But it was never discovered how far German Naziism was involved in the murder of the Austrian Chancellor, Dr. Dollfuss. For the first time the diary provides a clear and unequivocal answer to this question.

At Hess's one always meets interesting people. The most interesting was not Professor Haushofer but this Dr. Mader from Vienna, whom Hess has taken under his wing; because this young man saved a most compromising letter during the National Socialist riots in Vienna, in July, 1934; then he destroyed the letter before the police could catch him. In this letter Rudolf Hess had explained openly that if it came to a coup d'état, Dr. Dollfuss was not to survive under any circumstances. Planetta (*the man who shot Dollfuss and was hanged in Vienna for the murder*), who had seen this letter, had been chosen to carry out the death sentence passed by Hess on Dr. Dollfuss; he did his duty as a courageous National Socialist and died for it. But Dr. Mader smuggled the dangerous letter from the Chancellery in Vienna and destroyed it at once. I understand that Hess is grateful to him because of this.

Munich, Autumn, 1937

. . . Now the time begins again when I must read to him for hours, days. He doesn't sit still and rest, oh no, he wanders about uneasily and stares at himself in the mirror as if he were a stranger. But all this must not get me flustered. Sometimes, while

I read, he takes another book and starts to turn the pages; when I ask him whether I should stop, he says, no, no, go on, go on, slowly and clearly. Now I don't know what to think—can he read and listen at the same time or doesn't he hear me at all? He is now full of plans for the future, he wants to study architecture, "that is the only art that concerns me," he says. "I have often thought I could write short stories but as I could only compare myself with the very greatest, I gave up the idea. And painting is somehow too petty for me; and we don't live in an age in which painting can express anything." Then he stepped again in front of the mirror, looked at his tongue which is always yellowish, probably because he has an enlarged liver, and said: "I want to learn to speak English but no one must know. I consider languages dangerous, the most dangerous virus for spreading political and cultural power. Only when the whole of Europe speaks German, will I have achieved my aims . . ." I was deadly tired for it was about seven o'clock and I'd been reading aloud since three in the morning. The lines danced in front of my eyes, and *he* still did not want to go to sleep. At last he went into the bathroom but I had to go on reading, even when he started his usual morning exercises, those knee-bends and arm-stretchings. Only then did he say: "Enough for to-day. Interesting." And he sat down on the sofa and rubbed his hands, stroked Rex and finally said: "Get into the bath, then we'll go for a walk." He always watches me when I take a bath and fortunately I feel that he can have never enough of it. He likes especially to watch me while I do exercises—and I must do always those which he has devised for me though I've been doing physical culture long enough myself. "You are very elastic," he says now and then with appreciation, always using the same words, and then he calls my attention to various details. He certainly knows my body far better than I do.

The " solution " of the Austrian problem—as Eva Braun's notes prove beyond doubt—had been planned by Hitler and Rudolf Hess for September or November, 1937. Papen, the German Ambassador in Vienna and his military attaché also considered this date as suitable. Here is the almost incredible explanation why Hitler postponed his action against Austria.

November, 1937

Sunday. Gloomy, dull weather. Professor W., our court astrologer, and I, waited to-day for a long time for *him*. In the end he came at six instead of three. At first he did not want to see W. at all and told me to send him away, there were important things going on, the Anschluss of Austria and Germany were possibly about to take place. Hess intended to carry out the plan he had worked out with Papen and Tafs (*the chief organiser of the Nazi Party in Germany*) in the summer. A. was very excited. I hadn't seen him in such a state for a long time and I noticed naturally that he wanted to be alone with me. He needed an opportunity to talk, he wanted tenderness. But then he decided to see W. He greeted him gaily and asked: " What are the stars saying?" "Good things, my Fuehrer, for the time being, only good things," replied the old man. " What does it mean: for the time being? Do they foretell something bad for later?" Old W. is not easily flustered; he talks a bit like a Jew, perhaps he even has Jewish blood. " The coming year of 1938, my Fuehrer, will be the greatest year in your life, that's what I mean." "Does it mean that after 1938 my star will start to decline?" "A star like yours, my Fuehrer, does not decline so easily—either it goes on shining brightly or it is extinguished." "Is it to be extinguished?" "My Fuehrer," said old W., who was now feeling very alarmed, for the whole conversation had become alarming, "not even an astrologer should talk of such matters. I have prepared your horoscope for 1938 and can only repeat: it will be the greatest year in your life history." "Not 1937?" "No, 1938." He said goodbye to the old man and then said to me: "We don't have to talk much, W. and I, we understand each other at once. He advises me against starting the Austrian business now. Papen and Muff" (*General Muff, the German military attaché in Vienna*) "have been here. According to them this is the best month. I can and must not make my Austrian party comrades wait another winter." He walked up and down nervously and then drummed with his fingers on the table. "It is no disgrace to depend on the stars if one leads a great Reich," he said, more to himself than to me. "This Austrian question excites me more than any other experience in my life. Sometimes I feel as if something warned me against it and the old chap also pointed to the danger. It seems to me as if I were touching something . . . forbidden. These magic signs have always helped me, they are stronger than any intelligence or reason of state. And yet I'll solve the Austrian problem. Even if everything were at stake a hundred times, everything for the Reich. The Reich can only exist together with Austria or it must go under." He was so excited—I had never seen him like this before. And later he made love to me . . . and I felt he loved me very much . . .

Baldur von Shirach who, in November, 1937, was still the Fuehrer of the German youth, had the impudence to maintain in front of the Nuremberg court that he had shown the German people the road to Goethe. But how does this " spiritual leader " appear in the circle of his friends and companions?

November, 1937

The party at the H.'s was more amusing than usual. Of course, Erna is always a bit condescending with me, as if I were her employee—but she is such a primitive creature, though very gay. If Baldur is there (*Baldur von Schirach the leader of German youth, later Gauleiter of Vienna*) I always amuse myself with him. He is my special " flirt "—of course, quite innocent; but I act Gretchen a little and he acts Faust and we both like it a lot. He tries to seduce me and I resist. No one can take it amiss that he kissed me in the hall and I think poor Faust was quite frightened when he committed this high treason. But, of course, it has to stop. There was some silly goose from the Hitler Youth who clung to him the whole evening to everybody's amusement. By midnight he was completely sozzled and the gang locked him and the girl into a room on the first floor where they were supposed to get " married " with universal blessing. We went downstairs again. When we " released " them about six o'clock, the girl rushed silently and tearfully from the room. The poor boy who had apparently eaten and drunk too much had made an indescribable mess of it. Erna said, it must have been a sad wedding night. B. soon recovered, however. We washed him and cleaned him up as if he were a horse until he was quite awake. Then he wanted to start drinking again but all of us protested. As he seemed to be sober enough,

I gave him permission to take me home. He is always the same bad-mannered child. Naturally he tried to make passes again and when I stopped him firmly, he became terribly dejected and said that the only thing that still interested him in life was to own a brothel in Budapest. I was almost speechless. You never know with him whether he's joking or means something seriously. In any case these are not suitable future plans for the Reichleader of Youth. I told him that openly. But he clung to his idea and told me, just wait and see, I'll get what I want. And that all this was far more sensible than this stupid idea about the Mozart Festival in Salzburg which Axmann planned, and which would put the Bayreuth Festivals into the shade. Well, I let him talk for he was beginning to babble all sorts of confused nonsense and finally *I* took *him* home and not the other way round.

Obersalzberg, December, 1937, Sunday

. . . Yesterday we had a charming evening, quite domesticated, which was according to my taste. We ate alone, then he took his evening foot-bath—as usual, fully dressed, sitting on the edge of the bath-tub. The water had all sorts of bath-salts in it to resemble sea-water as much as possible. I read to him from an old book about Alexander the Great. It was so peaceful and quiet. *He* was in a good temper and patted my shoulder again and again. And his face was so relaxed, his cheeks were hanging down . . . I cannot get used to these sudden changes of his sometime uncanny character . . .

In the notes which Admiral Raeder wrote down during his imprisonment in Soviet Russia, the so-called " Moscow Memorandum " he says of Goering: " The only man who behaved tactlessly again and again and was openly bad-mannered to Hitler, was Goering. But Hitler accepted this behaviour without any objections."

December, 1937

I don't know why Hermann must always play the court jester. Sometimes I feel that he prepares some special turn every time he comes to visit Adi. Of course, under the mask of the court jester he can allow himself far more than any other man. I believe, on the whole, he is the only one whom *he* takes seriously. Of Goebbels *he* told me recently: " He is finished. Always tripping over petticoats . . . If I ever get rid of any of my old party comrades, it will be Goebbels."

Goering has the unpleasant habit of pinching my posterior. I don't know where he got the habit, probably he learned it from his intercourse with waitresses. And then he always greets me with the same question: " Am I getting a kiss to-day?"—The Fuehrer has a revulsion against baths, he only bathes once a month; at the same time he is amazingly clean, he washes himself thoroughly every day. I believe he does not like to be naked and I don't know whether anyone has ever seen him in swimming trunks. But he does not object to *my* nudity. He sits there, completely dressed and always very neat, the neatest man I know, and watches me carefully as if he wanted to memorise every movement. Because he does not like to take baths himself, he does not want to hear jokes about this subject. Goering knows that quite well. But recently Hermann has become as cheeky as if he felt himself wholly an equal.

And Adolf accepts this—to my annoyance. The joke about the Jew who forgot his undervest in the bath and discovered it only after a year when he wanted to bathe again, was much too ancient and stupid; Adi looked as if he was about to become very unpleasant. But in the end he said nothing.

Winter, 1937

Yesterday the house was full of guests; though most of them had to return after an early dinner to Berchtesgaden. A few stayed, among them Leni. We did not see each other. She does not know that we are meeting here to-day. I don't know at all what's going on down below. I was forbidden to go down. I must wait in the bed-room, in a nightgown, until he comes. I wonder whether they are performing the nude dances down below of which they always talk and which I must never attend because I am "a little girl" and "the secret queen"? I must always think of Leni. She is always abusing people, *he* told me, and that I don't like at all. But somehow he is fascinated by her and I don't know whether she won't oust me one day.—He has just left me. Oh no, she won't take my place!

Munich, December, 1937

Thank God, the period with the Bleuer book is ended at last! For two weeks I went to him every day in the Braunes Haus and took him a copy of the "Textbook of Psychiatry" in the locked briefcase to which only he has the key; after he read in it and made notes, I took it away again. "No one must find the book in my room," he said, "it might lead to wrong conclusions." And he looked at me in a way which made me quite frightened. He couldn't possibly think that he is . . . mad? When he gave me that lecture about " genius and madness," I felt cold shivers running down my back. He always becomes so mysterious and secretive and that's uncanny. " A genius," he said, " lives in a different mental dimension from the normal being, but he must possess the possibility of returning into the mental sphere of a normal being. If he loses the road back, then he appears mad to normal people, like Hoelderlin or Nero." Then he went on to explain that most geniuses do not know that there is a limit, a border-line, and they are not in danger. " But *I* know it," he said, " just as Shakespeare knew it, as his sonnets prove. Shakespeare could take a walk between the two dimensions. He was a gentle and amiable man. But I am violent. I'll simply burst open the border-lines . . ." His eyes glowed . . . as if he were feverish. It was uncanny and I was happy when I got outside at last. I made a note of everything so that I could ask Dr. M. what all this talk really meant.

1938

1938 was the year of Austria's annexation and the addition of the Sudetenland to Germany; the year of the most dramatic international conference of world history in Munich; the year in which war approached with giant strides. It was the year in which Adolf Hitler, as he told Eva Braun, began to worry about " *his* war." While preparations were being made for the subjugation of Austria, the Fuehrer, this " Spartan man," also occupied himself by giving his mistress advice in make-up and sending her " orders " about her underwear. The escapades of the different Gauleiters also reached a climax in the last year of peace—at least according to Eva Braun's *chronique scandaleuse.*

January, 1938

Wednesday evening. I am really touched for *he,* this Spartan man, had a long discussion with a cosmetician on how a woman can best retain her youth and beauty. The creams which he sent me seem to be really good. In any case, I am going to use them if only because they come from him. He took the trouble to write to me personally and draw up all this advice! Twice a week a night-pack of raw, fresh veal and once a week a bath in warm olive oil. The most important, he says, are breasts and hips. He is certainly an expert in these things, that's the artist in him. I must watch my hips, he says; all the rest is fine, legs and tummy do not show age, that is well-known. I must have no massage; he put a definite ban on that; he dislikes paraffin packings, too, for he once heard something about cancer being caused by them. Dr. M. says that it's pure nonsense, but what can I do? Who has the courage to explain to him that something he has said is utter nonsense? Also, I don't believe M., he is such a perfect cynic, he experiments with all of us as if we were guinea-pigs. I have discovered, by the way, that A. is always right in everything, even when it sometimes looks anything but right. That, of course, is why people believe in him. And if he declared to-morrow that the sun moves round the earth and not the other way round, the whole of Germany would believe it at once. I certainly took some time to get used to the chamois leather underwear which he wanted me to wear! To-day I can only say that it's the finest, most pleasant and comfortable underwear in the world. It never sticks, and knickers and shifts feel like velvet on the skin. Much finer than silk. It's the same with night-gowns. When I wore pyjamas, I couldn't imagine how comfortable night-gowns could be. As he always wears a night-shirt, I changed to night-gowns at his wish; he loves them with lots of ribbons which he can loosen and then let them fall down slowly. I like it, too.

The Austrian tragedy which started in February with the Hitler-Schuschnigg meeting, was preceded on the morning of the decisive day of negotiations by a farcical situation.

Obersalzberg, 1938

The fact that Machek, the chiropodist, did not arrive in time on the Obersalzberg, as ordered, certainly made the position of Schuschnigg worse in the negotiations about Austria. A. raged because Machek hadn't come and he had difficulty in putting on his boots. Machek's arrival was long overdue. Already in the morning A. was in such a bad temper that I feared the worst. As I heard later, Schuschnigg was treated very badly indeed and A. quickened the tempo of the negotiations and went further than he had intended—all because he suffered from his corns, which made him angry and rude. The main thing is that everything went well. Machek was involved in a motor accident.

Streicher was Gauleiter of Franconia, bearer of the so-called "flag of blood," Hitler's anti-semitic pioneer, editor of the "Stürmer." In 1939 he was relieved of his post by Hitler. The reason was peculiar. Streicher had sent about a dozen girls to a Nuremberg gynæcologist for illegal operations to be performed on them; he tried to create the impression that they were all Hitler's mistresses. At the Nuremberg Trials, Streicher declared that the Jews were nowhere better off or more unmolested than in his Gau!

June, 1938

Julius Streicher invited us to his " model estate." We drove for about forty-five minutes and then inspected the farm with the mansion. The latter is most nobly furnished, with red plush furniture, tapestries and thick carpets, a cross between Obersalzberg and Karinhall. Streicher took his guests round and played the farmer. I was tired and in any case I am not a bit interested in farms. Garner told me that the main amusement was anyhow most nauseating. Streicher's favourite game is to keep a large ox without water for days in the greatest heat—and then give him so much to drink that he bursts. That tickles his sense of humour! K. maintains Streicher is fundamentally a jovial fellow and not at all the Jew-baiter he pretends to be. He is proud that some Jews still have shops in Nuremberg. True, the wives and daughters of these Jews must visit Streicher once a week at his Gauleiter's office where they are forced to walk naked in front of him. I wonder whether it's true?—In the evening there was naturally a lot of drinking. It ended by Streicher making his appearance on the staircase at midnight, dressed in nothing but his Gauleiter's cap, gloves, high-boots and sword.

In order to remove intellectual resistance to its doctrines, National Socialism gradually and systematically destroyed the foundations of religion, free science and law. A single instance which Eva Braun describes quite objectively, provides a picture how far the destruction of legal authority had progressed in Germany as early as 1938.

Munich, 1938

In the afternoon the Chief Public Prosecutor Merck called; he insisted on seeing me alone. Frau Almas was terribly curious and would not leave the room. She smelt scandal—and she was right. At last I got her out. Merck hummed and hawed and asked me ten times whether he could count on my discretion. He has a moustache just like Adolf's — how different it looks on another man's face! After I had told him ten times that I can be as silent as the grave, he was still beating about the bush and fingering his collar which had become quite limp while he sweated. At last he came clean with the following story: For the last three months he had had a report of the "Elisabeth Heim" maternity home on his desk. A girl of twelve-and-a-half had had a child; according to German law this had to be reported to the state attorney. The girl told the doctor that she had been raped by the Gauleiter. The Chief Public Prosecutor said immediately that this was of course nonsense; at least ninety per cent. of the women who complained of rape, hadn't been raped at all according to forensic practice. Here, too, the facts were not clear and probably the girl hadn't resisted sufficiently—that was all. But the doctor insisted on legal action; the President of the Court would not interfere—and it was now up to him, Chief Public Prosecutor Merck, to bring the charge—which was unthinkable and impossible. He dropped a hint to the doctor through the police that he might lose his job; but the prioress of the nuns visited the Police Chief and protested. I wanted to know who the girl was and how Wagner had met her. "That's just it," said Merck, "she is the daughter of a maid in Wagner's house." (*Wagner was the Gauleiter of Munich.*) I didn't know what to do in this whole affair. Speak to Wagner, the Public Prosecutor asked me, and looked at me like a bunny-rabbit sentenced to slaughter. Fundamentally the whole matter was simple. He had already made his plans: the maternity home is to be cleared of the nuns who are to be replaced by National Socialist nurses, the obstinate doctor will be sent to Dachau and the twelve-year-old mother to the closed department of the clinic for nervous diseases. I wanted to have nothing to do with the matter. The question of how the fat W. had "raped" a twelve-year-old girl was really most ridiculous. Why shouldn't he face a little unpleasantness in exchange for such undeserved pleasures? I wanted to make it clear to Merck that my intervention might lead to more painful consequences than the legal charges, when Frau Almas rushed in. She had overheard everything by "accident" in the next room. She offered to "arrange" the matter. She went straight to Frau Wagner. At nine o'clock she told me that the Elisabeth Maternity Home had been closed by the Gestapo.

SS. Chief Group Leader Kaltenbrunner was head of the police in Austria; after Heydrich's assassination he became Himmler's deputy in command of the SS.

Munich, summer, 1938

. . . Kaltenbrunner, who as Hoffman says is a coming man and, more important still, the only human being whom Himmler fears, took me through all the rooms of the *Braunes Haus*. That is, I was really his guide. It was really beautifully decorated, Gothic rooms, Renaissance halls, rococo . . . everything together; then the coats-of-arms of the German cities and provinces, silk, brocade and gold; it made you feel quite dazzled. K. was not so much impressed. He measures everything with a cold, non-committal look. Again and again he asked: "what does this cost?" and again "how much was this?" I don't know the figures and replied to him rather sharply. He introduced a man called Globotschnik (*Gauleiter of Vienna*), a good-looking man. K. suddenly woke up and said: "I congratulate your discriminating eyes. This man has already committed murder for the sake of the Party." My brain seemed to stand still. "He killed a jeweller in Vienna during our illegal days." I couldn't understand how this concerned the Party. "Well don't you understand—the jewels, the booty, he handed over to finance the Party . . ."

A Jewish jeweller was murdered and robbed in Vienna, in July, 1937. SS. circles were suspected of the crime, but it was never wholly cleared up.

Friday, summer, 1938

This summer might have been the loveliest in my life but now, again, everything is spoiled. I was so happy all these days at the Tegernsee, lying in the meadows, staring into the sky and knowing that *he* loved me. Can you expect more of life? But my whole relationship with him is under a baleful star, he cannot leave me in peace, again and again he makes me uneasy in some way or other. I am proud that he told me all this in addition to what he already confessed in Vienna, but at the same time it oppresses me in an indescribable way. And I feel physically . . . anxious. But I know, on the other hand, that he can never, never leave me now. He brought me a thin, funny platinum ring with an inscription in relief: "Until death do us part." He looked at me in such a way that my mouth went dry, and then he told me the story no one has ever heard. "In order that you may know that I am bound to you as to no one else."

Saturday

I couldn't go on; my sister came and she always laughs at the diary, but thank God she is not curious and doesn't know at all what I am doing. I am still terribly sad and the wonderful weather makes me quite melancholy; I must get myself a text-book on venereal diseases in order to see the whole story clearer and understand it better. Well, this is how it happened: on his seventeenth birthday he went for a walk on the Ring with a friend. A. wanted to borrow some money from him for a big illustrated book on the history of art, which he had bought rather rashly without having enough to pay for it. His friend agreed but only on condition that A. would first go with him to the girls. It was his first experience. First they decided to drink something . . . Truly, this is a terrible story of what happens to such innocent young people. He had never tasted a drop of alcohol. His friend gave him a whole glassful of Schnapps—in order to get him in the mood, and to remove his inhibitions. From the moment when they stepped into the street again, he knew nothing of what happened, he told me. He woke up many hours later in the grey dawn, lying in a wide, dirty French bed; at his side there was sleeping a fat, naked, no longer young female. He was sick. The woman woke up, grinned, yawned and said: " It's all paid for!" and turned on her side to go on sleeping. " Don't make a mess!" she told him, and as he opened the door: " 'Bye, bye, little boy!" He told me he had never dressed so quickly in his life. Once in the street his only idea was to get to a doctor at once. But it was only seven o'clock, the front-doors of the houses were still locked. He wandered around until he found a doctor who started his surgery at eight o'clock and asked him to examine him. The doctor was a callous, tired old physician who said: " Don't worry, probably nothing happened," and gave him a disinfectant. He must have been very unsympathetic. Instead of reassuring the seventeen-year-old boy, he treated him as if he were a pimp who had made a night of it. I believe Adolf's stubborn dislike of doctors and of medicine in general springs from this experience. In any case, the doctor told him that the incubation period was twenty-one days and that there was nothing to do but to wait calmly until it was up. The idiot! How can one wait calmly under such a cloud! But the three weeks passed somehow and no symptoms appeared. On the 23rd day he again visited the doctor who examined him and congratulated him: there was nothing, absolutely nothing. A. went home, full of the best resolutions in the world, swearing an oath that he would never again visit a prostitute and never again drink to excess. In the middle of the night he awoke—he had a queer feeling of tension. Half mad with fear he jumped out of bed, dressed quickly and rushed to the doctor. It was two o'clock in the morning. He woke the doctor and forced him to examine him again. The result was unfortunately only too evident. I can imagine how desperate he was over this sad story.

But he possessed the iron will not to drink a drop of alcohol for twenty years. His diet, at which many people mock, also began in those days and he has never deviated from it, not even during the world war. That also explains his complete cure.

Monday

Read in these nasty, disgusting books the whole day. Seeing those illustrations, one gets frightened and disinclined to touch any human being. In any case, now I know enough. Naturally it was as he told me, a hundred per cent. curable. But I now understand those attacks of violent irritability and deepest depression. I don't know whether he regretted having told me; for yesterday he said that we must never, never again talk about it. I am certainly in agreement. But it will give me food for thought all my life. The same thing is said about Napoleon and also Mussolini. Probably it is like a childhood illness with all men—but no one ever finds out when it happens to someone unimportant.

Summer, 1938, Obersalzberg

The friendliest evening for a long time. The whole day messages from Berlin. Frick was here, Frank and Rosenberg, each of them with a different plan to conquer Europe. Each of them shows that he wants to take part in the birth of Greater Germany and what follows afterwards. In the evening it was quiet. Adi and Hess sat in front of the card-table placed at an angle and talked softly. The table was strongly illuminated. I sat in the corner, knitting and dreaming in the semi-darkness. The gramophone played "Tristan and Isolde" and now and then A. whistled the melodies. Hess was wearing shorts. Sitting side by side, they looked like brothers. A. kept on touching Hess's bare knees and legs, while pointing to the map with his other hand. Then they looked so strangely at each other. I began to have funny ideas. I must really find out how it was with Roehm.

Munich, Summer, 1938

It rains all the time. Had a funny experience Met the Fuehrer by accident, walking alone in the rain under a huge umbrella in the Kauffingerstrasse. As I had met and recognised him, he told me he did this often; he loved to take long walks in the rain, alone under his umbrella. I didn't know of this at all. He doesn't want people to find out.

On the 1st October, 1938, the press of the whole world celebrated the Munich Agreement as a victory for peace. Only the German papers displayed a rigid and reserved attitude and emphasised that Munich had supplied proof of the rightness of the Axis policy. In the House of Commons there were scenes of wild celebration. Winston Churchill delivered his great speech to which little attention was paid, pointing out that Britain had to choose between war and shame; she had chosen disgrace and should have war as well. What was Adolf Hitler saying?

October, 1938, Munich

Three of the " Big Four " have left—the fourth, undoubtedly the Greatest, has remained. For three hours *he* told me of the negotiations. " Only now do I know how weak the West is," he said. " And now I'll make the war I need to carry out my ideas in the world. The difference between Mussolini and me becomes clearer and clearer: he worries for his peace—but I am worrying to get my war."

Munich, Autumn, 1938

. . . Always he wants to undress me himself! He has such nice strong hands which make me quite crazy . . . But he can't undress me because he is simply not skilful enough. Then he wants me to undress him—well, as far as he undresses. Everything is always complicated and must follow prescribed rules—and these I haven't learned yet. He says nothing, of course, only if I make a mistake, he becomes sad and then I feel like howling. I cannot bear his sadness, it's too heavy a burden for me, there is something mad and irrevocable about it.

1939

1939 was a tragic year for humanity. The tremendous tension of 1938 which lessened slightly after Munich, grew in a steep curve during 1939. The highlights were the incorporation of Czechoslovakia in the Reich, the attack on Poland, and the British and French declarations of war. These were surpassed by the quick advance of the German troops in the east, showing that the Nazis, for the time being at least, were invincible. Eva Braun's diary is little influenced by these world-shattering events. She only notices the things immediately around her and is completely filled with her relationship to *him*; she is mainly concerned with gossip and petty jealousy, with personal and private problems and meetings with the SS-"aristocracy." 1939, however, brought the first sign of the beginning resistance to National Socialism in the attempt on Adolf Hitler's life during the " traditional gathering " in the Munich Hofbräuhaus. A man called Elssner was accused of this crime; according to official German reports he was arrested while trying to escape to Switzerland. Eva Braun provides significant information about the background of this attempted assassination.

Summer, 1939

I think he has now suddenly decided that I alone shall remain his " secret queen." Naturally he told *me* nothing about it but I have noticed that for some time Leni has been eliminated as a rival. Last night he told me, while his eyes roamed over my body: " You are the only one of whom I never tire, your body will always remain a mystery for me even if I look at it every day for a thousand years." I am not the loveliest woman he knows, he told me. I knew that, too. But I am supposed to hide the " mystery of eternal fascination." I am very happy about it. I don't know why he always remains dressed. I can be as nude as I want to be. It's always the same: " Aren't you too hot in those clothes?" Naturally I understand that and take them off. But *he* always remains fully dressed as if he were at some reception. He always feels cold but that isn't of course the reason why he keeps his clothes on. He is simply shy. I was, too, at the beginning, but that passes. But *he* always turns away when he undresses. And yet I never look at his body. It is his eyes which have a hypnotic power over me; when I am with him, I am a different being altogether. Especially when he is excited and has that carnivorous look in his eyes. For the first time we spoke seriously about Leni. Until now he has always only smiled when I tried to find out something. Now he says: "But she is a great artist and an important human being." What do I care—if she only leaves him alone? "As a woman I find her uninteresting," he maintains and now I believe him. "She has some attraction which appealed to me, not erotically, but because of her artistic vitality." They couldn't have been really intimate, though he gives her an awful lot of money. I asked him whether she had a beautiful body. "Yes," he said thoughtfully as if he had to recall it first, "she has a beautiful body, there is something Greek in her, she has a cold sensuality. She doesn't walk beautifully, she is not graceful and affectionate like you . . . it's all instinct. And that always repelled me." He was silent for a while, then he said: "She always wanted something, always. She is the most ambitious woman I know. Sometimes I was actually tempted to give her more and more power, to raise her higher and higher—in order to see where she would end with her unusual temperament. But I have no time to act the Sun King. I have other tasks to fulfill." As he mentioned her "cold sensuality," I remembered the passage, the only one he had underlined in Shakespeare; the one about not wanting to become a fan or a pair of bellows to cool the lustful gipsy-girl. That could only refer to her. I am still tortured by this: did he have an affair with her or not? Am I ever going to find out? Of course he does not want to give *me* any power, and I never want anything from him, and have never asked for any favours; I am probably the most convenient mistress for him.

August, 1939

I really could be jealous of H. (*Rudolph Hess*) if he wasn't such a nice person. Every time he comes,

Adolf smiles, puts his arms round his shoulder and walks up and down with him. I believe he tells Hess almost everything. To-day I asked H. how he came to the idea of incorporating Austria, for it was originally his idea, Adolf told me that himself. Hess replied that he was an anthropologist (whatever that meant) and continued: " First of all I wanted to gladden the Fuehrer's heart and give him his own country back; and secondly I knew from a different dimension that it was necessary and had to happen. Only the date was something we couldn't agree on with the Fuehrer. I wanted it to be September or November, 1937. I must have misunderstood my demon; the Fuehrer was right. March was exactly the right moment for it."

Summer, 1939

He came to the dressmaker with me, looked at everything but was silent like a stone. He says nothing and then later he reproaches me that I have chosen the wrong things. Then we drive out to the Nymphenburg Lake — there was a water festival there with naked dancing mermaids in Bengal lights. I noticed that he liked one of the girls and that he had her presented to him. She came up quite shamelessly in her transparent veils through all the people to the terrace and curtseyed most exquisitely. But he did not like this shamelessness; I know him. After all, he himself is terribly shy. I believe he would like to have an amorous adventure now and then; there are girls he likes but he is simply too helpless, too clumsy, I don't know how to put it. He simply doesn't know how to do these things. Probably he is afraid, the others would notice what he wants—only the girl wouldn't. He has no idea that girls realise such things very quickly; they know what he wants. But let's leave things as they are . . . His shyness must be connected with his bad experiences. It isn't easy with him. I now know his weaknesses as well, and I know what he wants and he always wants the same thing, always the same. Another woman might take him for an ordinary man and that would hardly satisfy him . . .

. . . But his reserve, his shyness is sometimes not normal. Thus it's always a problem for him if he is in company and wishes to go to the lavatory. He always has to fight inhibitions, I often notice it. If he were a girl, he would probably sink into the ground with shame. And yet when we are alone, there's this mad, wild flame and fury!

Mid-August, 1939

. . . H. (*Hess*) told me about Schuschnigg: "He was not the head of a government but simply a clerk in the chancellery. We could have dealt with him at any time. He had no friends either inside or outside the country. He also committed a decisive mistake by failing to give the order to shoot. He was too soft, too weak, too intellectual. An intellectual is always sentimental when it's a question of shedding blood. With his capitulation Sch. betrayed his country to all practical purposes. That's why he was allowed to stay alive—and produce children. The Fuehrer only punishes traitors who have betrayed *him*. The others he allows to live, for tactical reasons."

August, 1939, Tuesday

The rest on the Achensee was wonderful, the scent of the woods, the blue lake and the boats far out! I had my swimming suit in the suitcase and was too comfortable to unpack. Two roads were blocked at once and I could swim in the nude. *He* never bathes and I don't even know whether he can swim. He stretched himself on the shore and watched me. Then we lay on the moss and I let the sun dry me. He again wanted me to tell him of my childhood . Naturally he always wants to hear about certain things—but there aren't many of them in my life. I told him that I was fourteen when I started to compare my breasts with Rita's. I still laugh when I remember what silly kids we were. We used coffee-cups and I was always behind the others, they were far more developed. Rita consoled me and told me I would be more seductive! But what did we know of seduction then? Adolf, however, says it's always the girls who seduce the boys and usually the sisters are the first seducers. " I received my first erotic lessons from my sister," he said. " Girls mature much sooner than we do. I must have been twelve and my sister over thirteen when she wanted me to draw her in the nude. And she got what she wanted. In those days I was a wonderful draughtsman; if there hadn't been so much terrible trouble at home, perhaps I would be Europe's most important painter to-day. As far as the intensity of artistic feeling goes, no living man is my equal to-day—and I don't know whether it ever burned so strongly in any one except, perhaps, Kainz (*the famous Viennese actor*) and Goethe. And Goethe damped his own fire deliberately and arbitrarily. With me it might have burst out perhaps . . ."

Summer, 1939

. . . The afternoon on the Chiemsee was completely spoiled by the terrible fit of fury which suddenly overcame him, when he noticed I wasn't wearing any underclothes. But on such hot days I usually only wear a dress, like most women do. He roared at me in the street like a bull; he shouted that this was like a whore's behaviour and I was evidently quite depraved and degraded. I said nothing, only looked at him. Then he recovered his poise a bit and said: "It must never happen again! You might fall down . . . how would that look?" The whole evening he was bad-tempered, silent and irritable. I hurried to put on something at once and told him softly about it but I couldn't drive away his black temper; he only looked at me grimly and said: "Well, then."

The 1st September, 1939, shook Eva Braun a little. She cried. Not because war broke out but because "his voice sounded so strange" and she was probably overwhelmed by the final passage of his speech in which he declared he would not take off his uniform until the war ended; he would die if Germany did not achieve victory. Was she sensing her future fate? But on September 3 when France and Britain declared war, she ended her diary with the laconic remark: "All in all a boring autumn Sunday without any special events." The same spirit produced the following joke in Germany: in order to spare the Fuehrer any small irritations, a notice was put up in the commissionaire's room at the Reich Chancellery: "Further declarations of war to be handed in *here*."

September 1, 1939, evening

So the Polish campaign has begun. Heard his speech on the radio. I cried, his voice sounded so strange and I was afraid. At four he came suddenly and said: " Warsaw is in flames. Scum, Jews, German haters. In ten days the whole of Poland will be burning. Then, at least, they'll have their brains lit up. In ten years there won't be any Polish aristocracy: the formation of an *élite* will be made impossible in Poland. They have had several centuries to prove their worth in history. They already behaved badly to Napoleon. They won't be given any opportunity to do the same with me. They have been weighed and found wanting. The Russians colonised badly there. They tried it with deportations, dungeons, tortures. That was, of course, wrong, like all the Russians do; it was wrong in the higher sense and remains wrong. I'll do it differently. Not like the British, either. Their method of penetration is too slow. I have no time for that and who knows who follows me. Goering is a fathead, sentimental and cynical at the same time. My system in Poland will be drugs. The IG Farben can produce enough. Here, too, I am a pioneer. We need no intellectuals in agricultural districts but working slaves. The Poles are heavy drinkers. With a weekly rum ration of half-a-pint, adding the IG-drug, we have reduced in seven years, according to the Planck plan, the Polish standard of intelligence to the same uniform low level, have excluded the formulation of political ideas . . ." He was again as in former days and I told him of my anxiety during his speech in the Reichstag. He laughed but he was uneasy and he looked at me until I felt dizzy. I lay down and he came to me. At seven I stumbled through the newly blacked-out city to the Ufa cinema and saw " Prozess Deruga." The streets were full of people, all afraid. But not me, not any longer.

Sunday, September, 1939

The expected declarations of war have rolled in. He is in the East and makes war. All in all a boring autumn Sunday without any special news.

Late Autumn, Sunday, 1939

Was invited by Goebbels to a private luncheon. The room in which so many actresses have lost their virtue does not look depraved; more like a woman's. If you are alone with Joseph, there is no service. He has arranged that quite shrewdly for his many private adventures. He merely presses a button and the circular table disappears slowly through a trap-door in the floor. Down below they put the next course on it and the table appears again. Most practical, especially if you really want to be alone. It goes amazingly quickly. Someone told me that Joseph usually keeps the wine at his side and sometimes throws a bottle into the hole if, in his impatience, he finds the service too slow. That might be true for he treats his servants like Chinese slaves. But when I lunched with him, he behaved most correctly and made love to me as if I were a lady of sixty. Someone was said to have threatened him with jumping into the serving-hole if he did not leave her alone. In his place I wouldn't have believed her for it must be very uncomfortable—to jump in there. A young Viennese actress is said to have kept him at arm's length by confessing that she was in love with him; she said that was the reason why she couldn't yield to him for it would make her crazy, she would follow him everywhere and demand that he marry her. That, too, seems to have worked. With me he is either whimsical or very matter-of-fact.

Obersalzberg, 1939

At last Schaub and Wernegg have left with the body of the girl and the bedroom is being cleaned up. It is almost three o'clock—they'll all be asleep at the " Turkish " place. I am sure they'll rouse the whole house and everything will become public. I am quite exhausted and will try now to sleep, come what may; my nerves can't stand it any more. If only I can sleep. Down below the celebrations go on and the mother of this mad girl has no idea of the tragedy which has taken place up here. But I have known for months that this would end badly. Even if no one could expect that she would have the courage to put a bullet through her head in *his* bedroom of all places—using that heavy army revolver which blew half her head to smithereens.

Afternoon

The whole haunting business is over. *He*, too, has left and Baroness Elss thinks her daughter committed suicide at the " Turkish " place. Sch. told me she collapsed, fell like a stone, when they told her that her daughter was dead. Well, those louts probably didn't do it very tactfully, that isn't their way. Poor woman, I am sorry for her, she is a nice, cultured creature. But why did this unfortunate girl have to fall in love with *him* of all people? And fall in love so madly that she lost her head completely! She has been running after him for months. Then at last she met him and he caressed her fat little hands—and that was all. And the letters which she wrote to him—making Sch. and his friends roar with laughter! For *he*, of course, never saw these outpourings of passion. He knew, naturally, that she was in love with him but he considered it the innocent crush of a young flapper. Who could have suspected that this eighteen-year-old girl was not a flapper but a passionate woman, prepared to go to any lengths, do anything! When I came up, they had already bandaged the corpse; they used her own slip, covering her head, but she still looked horrible enough. And we had come up from dancing down below, full of laughter and music. There lay the poor little worm, not even winding itself. She was hardly a woman, a very fragile body, lying there on the bed, with the turban-like bandage on her head, she almost seemed alive. The clothes were scattered on the floor; she must have been a little drunk, for they were thrown in all directions. Sch. simply wanted to pack her into a trunk and take her away, anywhere. He always believes in the simplest solutions. But we didn't want to do that—the baroness wasn't supposed to know—and if possible the other guests should also be kept in the dark. W. immediately told *him*, but luckily he did not come upstairs. I, too, came only by accident into the room; and *he* raged when he heard that I had seen her; I was sorry, too, but no one could help it. He ordered himself, without coming upstairs, that she should be dressed normally, and that they should put her into the car and take her to the " Turkish " place. Then engage a room there; someone to fire a shot inside and pretend to be the first to enter, shouting that someone has committed suicide. It wasn't simple to dress her. They say corpses are quite stiff, but she was completely soft and the head and arms were hanging down. In the end the bandage burst and that was the most horrible thing. I almost fainted and wanted to rush out but then I thought that if I left the men alone with her, God knows what they would have done. It was quite impossible to put on her shoes—I gave her a pair of my slippers so that her feet shouldn't get soiled in the car. Finally we put the fur coat round her and the two men carried her out the back way, down the dimly-lit staircase into the courtyard. They had to get the car . . . in the meantime everything had become soaked with blood again, it was pretty horrible. But at last she was in the car. It was cold and the moon shone above the icy, snowy peaks as they drove away with their sad burden. I went upstairs and found her hysterical farewell letter under the bed, in which she wrote that she could not live without *him* and she was happy at least to die close to him—and it was a pity that she was not a man for then she could have sacrificed her life in fighting his enemies. Alas, this isn't the only case. More and more women go mad because of him, they are after him like devils and many have lost their heads and threatened suicide. *She* did not threaten anything of the kind, she only did it. *He* was quite calm; a human life means nothing to him, not even when it is destroyed close to him. " Most of these crazy creatures," he told me in the morning after he had dealt with the police officer who came up from Berchtesgaden with his report, " have only the ambition to act a little Pompadour. But you are nothing but a loving woman . . ." He caressed my hair. I felt proud.

Berlin, 1939

All the well-informed are talking about the divorce suit which young Erkelet started against his wife, accusing her of adultery with Goebbels. He offers proof in some letters which are certainly outspoken. A funny story; and this time serious for Goebbels because he has for once found a judge who is willing to defy even the Gauleiter of Berlin. Himmler refuses to help him because he is glad—he hopes that the story may become public and if possible reach the ears of the Fuehrer. H. knows that the Fuehrer has only recently told Joseph off in very strong terms and has warned him "for the last time" to stop "this damned chasing of women." But this twisted little devil can't give it up. If she loses the case, the foreign press is certain to get hold of it and then there will be plenty of publicity and—alas for poor Joseph! Thierack (*later Minister of Justice in the Reich*) is said to have told G.: "Don't worry. Follow the example of Napoleon. When a mistress of his youth threatened to publish his love letters, he answered: Publish them, my dear, and then go to hell!" But G. didn't seem to find any comfort in this advice. He is certainly right. In those days morals were different; and the little burgher who sets the tone in Germany to-day doesn't like any funny business if the honour of a married woman is involved.

Berlin, 1939, in the evening

Was invited to a meal at G. (*Goebbels*) in the Promi (*Ministry of Propaganda*). I couldn't help myself and directed the conversation to his "case" in order to tease him a little. But he seems to be rather gloomy and swears eternal revenge against the Erkelets, the judge and the public prosecutor handling this idiotic affair. During the meal he confessed, rather hesitatingly, that the affair was especially painful because he had had an affair in the old, old days with Frau Erkelet's mother; this is known to her son-in-law and might easily come out at the trial. And it would look as if he had had an affair with his own daughter! I can't imagine that any German judge of 1939 would permit even a shadow of such a charge to be raised against the Gauleiter of Berlin. But G. has a different opinion.

" The judge, of course, is an enemy of the state," he said, " or he wouldn't treat the case with this idiotic objectivity. For him it's a gala occasion." I wanted to know whether Frau E. was really his daughter. " No idea," he said. " She certainly doesn't look like me. Nor would I be bothered about such a thing—on the contrary. Are you shocked? You don't seem to know Greek history; for the ancient man it represented a special attraction to sleep with his daughter. The most important Greeks were of the opinion that the best results of cross-breeding were produced by the relationship of father and daughter, mother and son. They also agreed that the intensity of emotion is unsurpassed in these cases. Only later was it maintained that incest led to degeneration and especially to intellectual decay but I am not sure that this is true." I would have liked to ask what his personal experiences were about the " intensity of emotion " in such cases but didn't dare; one doesn't know where such a conversation might lead with him; this fellow almost sparkles with sex.

Berlin, 1939

The Erkelets' lawsuit has been stopped after another judge had taken over the case. The first judge was killed in an attack by young SA-men in Dahlem. His wife is in hospital with serious wounds and is unlikely to recover according to Keppler (*Secretary of State in the Foreign Office*) for she is being treated by SA-doctors. Young E. fled to Munich; it is said that his wife has once again become G.'s mistress. Thus this difficult matter has been settled automatically and Joseph is able to sample the highest intensity of emotion until his " next great love."

Himmler called A.'s attention to the fact that in

those circles hostile to the state they are now reading Jakob Burckhardt's letters and recommend them to others; he asked whether the book should be banned or at least withdrawn from the book trade. Adi saw the passages to which objection was raised; he smiled in his cold and superior way. The main objection seems to be to Jakob Burckhardt's prophecy of the coming of the *terrible simplificateurs*, the terrible simplifiers, altering the world and shaking it to its foundations. "If my enemies can think of nothing better to bring up against me, they must be a miserable crowd," said A. "Burckhardt is right, by the way: I am a terrible simplifier. In reality I am the only one who will shake the world to its foundations and alter it. The book must not be banned; it should be well distributed. Burckhardt's prophecy is brilliant propaganda for National Socialism." And he told Himmler: "You must remember —not everything that is said against me is necessarily evil and false nor must everything be banned."

The German propaganda steadfastly maintained that the attempt to assassinate Hitler in November, 1939, at Munich had been organized by the Intelligence Service and was paid by it. Hitler himself, in his New Year appeal to the Party on January 1st, 1940, referred clearly to this attempt when he said that "the Jewish-Capitalistic World-Enemy" was now bent on exterminating his person, i.e., National Socialism. But Heydrich, who himself later became the victim of the avengers in Czechoslovakia, looked at it differently.

1939

Had a long talk with Heydrich about the Fuehrer. H. maintains that if there is any dissatisfaction with *him*, it originates in Austria. There is less "inner opposition" in Poland than in the Ostmark where to-day even the party members are turning against the "Old Reich Germans" and judge everything that is ordered from Berlin with a jaundiced eye. There have been incidents with Austrian soldiers but these have been stopped at once. It is significant, said H., that the clues found after the bomb-attempt in Munich pointed to Austria where K. (*SS-Chief Group Leader Kaltenbrunner*) has now gone on a mission to uncover the plot. So it was a plot? Of course it was—and the clues lead, as he said, to Kufstein, Vienna and Wiener Neustadt. I asked whether the man who was arrested could not be made to speak? H. became very gloomy and said: "First of all, we don't know at all whether *he* is the guilty person; secondly, he died in our hands because he could stand less than our idiotic doctors thought; and thirdly I think he knew little, for he was only a subordinate tool, if he had anything to do with it—one entrusted with the technical details"

1940

The year in which German troops conquered Norway, Denmark, Holland, Belgium and France, the year of the *Blitzkrieg*, the climax of German armed might in Europe. For Eva Braun its main importance was a love affair in which she indulged but which Hitler liquidated with his usual ruthlessness. She took part in his triumphant entry into Berlin, and found it "charming," but fundamentally she was more concerned as to whether she had to share Adolf Hitler's favours with Leni Riefenstahl or not. Adolf Hitler himself, from Eva Braun's point of view, seemed mainly to be occupied with the anatomical problems which played such an important part in the marriage of Marie Antoinette and Louis XVI. He also planned a book of memoirs for after the war, and intended to call it "My collected broken promises." But even Eva Braun felt that this plan could not be taken too seriously.

Berlin, June, 1940

Various things have now become clear to me. I am reading Stefan Zweig's "Marie Antoinette." Of course, it's strictly forbidden; Jewish smut. And I don't like it. A. (*Frau Almas*) lent it to me. She has, of course, a whole drawerful of Jewish books; but she does not read them, she only lends them out. One day they'll find her out and then God help her. I don't know whether they won't shorten her by a head. Not long ago someone was beheaded because he declared in the *Englischer Garten,* that war was madness and that millions had fallen in France. But this isn't supposed to be true. *He* told me that the soldiers on the battlefields shouted his name until they became hoarse and could not speak. I believe he has never been as popular as he is now. But why does one see so many red notices on the walls? Thorak called my attention to them and I counted them: seven in a week in Munich. "National Treason!"

The frivolity with which Adolf Hitler estimates the length of the war in the letter mentioned in the following passage seems almost incredible. And yet other witnesses give the same report of the ideas that ruled the Fuehrer's H.Q. in those days. Very characteristic of the relations between the SS-leaders and the generals is the conversation with Goebbels.

Apparently Eva Braun has still heard nothing about the Fritsch-crisis; gaily she assumed that the naughty general committed suicide because he was not fêted sufficiently.

Munich, June, 1940

Returned from Berlin yesterday by plane. *He* is in the West. He wrote me that the war was nearing its end and that a new life is going to begin for me, too. What does he mean? Already in May he dropped a mysterious hint. Now, before he left, he told me again, as so often before: "You are the secret queen." I don't dare to think what this might mean. In the newsreels I saw the first battles in the West. Break-through all along the line. *He* made the plans all himself and I believe that's the only thing he is proud of. The Polish campaign was something different. It is said to have been planned by this General Fritsch who afterwards committed suicide in Warsaw because he wasn't sufficiently celebrated. Dr. Goebbels told me just recently in Berlin that the pride of the generals must be broken. The Fuehrer is the greatest military genius of all times and the German general staff dares to smile at him. I don't believe that G. only told me this so that I should repeat it to *him*. He especially abused this Herr von Reichenau who is supposed to have said: "How fortunate that the Fuehrer was a corporal or he would be even more common." But this sounds more like Goebbels, who can't stand von Reichenau. If I tell Adolf such stories, he grows pale. He cannot bear jokes being made about him and yet he always wants to know if any are made.

I have now finished "Marie Antoinette." It was really very interesting, for me, personally. I must try to find out more about Louis XVi. I read the chapter about his marriage three times. I have learned several things from it. It is very instructive for me to discover that other historical figures have suffered from a similar trouble to *his*. I always wanted information about anatomical peculiarities, but didn't know how to do it. Could I borrow a text-book of anatomy from the University Library? I won't have peace until I know exactly how it is. Stefan Zweig describes the troubles of his King most dramatically. That must have been beyond doubt a more serious case than ours. I am not sure whether I shouldn't ask Dr. M. But I don't dare to do it for this cynical fellow naturally understands at once and that would be terrible, unthinkable. I am not even certain whether *he* knows his " case," whether he knows that this isn't normal. Of the three women he has loved, I am sure that not one told him if they noticed it at all. With us things are different, of course, from Louis XVI. *They* wanted to have children at all costs. I don't understand why Marie Antoinette wanted to become pregnant. I wouldn't like that. Then she wrote letters to her mother—striking a balance like a book-keeper. She is, on the whole, unsympathetic, but the King is nice. I'd very much like to know whether it's because of his anatomical resemblance to Louis or simply because he is so inclined, that A. likes this kind of love-making. They say it's terribly unhealthy. It certainly does me no good. I become nervous and hysterical; I am only calm, anyhow, when I am not with him.

Munich, July 2nd, 1940, night

A party at Hoffmann's. The " guest of honour " was Himmler. They must always have a guest-of-honour. Late in the evening they even dared to toast him and call him "*Reich-Heinrich.*" I was told not to drink much and therefore wasn't very gay—

but the others became completely sozzled, one after the other. In the end even the servants drank French champagne from the buckets. The nicest person there was a young painter from Vienna. He is going to call soon . . . Oh well!

Munich, July, 1940, evening

Kurt was here for four hours; I am afraid I've got it badly. We sat on the sofa and drank tea and talked, talked, talked. That is, we didn't really drink tea. It had grown cold in the cups and in the pot. He told me of Vienna and of his painting. Naturally he has no idea who I am. The Viennese never know these things. It is as if Vienna wasn't in the Reich at all. Kurt rails against Bürckel and the Nazis who are said to have ruined Vienna; he abuses Globotschnigg and everybody else. But he does this very nicely, in the Viennese dialect. He asked me whether I was a Nazi. I asked *him* how he got to the Hoffmanns with these views of his, and he confessed that Frau Hoffmann had taken a fancy to him. I can understand. He is tall and dark, his black eyebrows have grown together, his hair is nice and his lips make me hungry. There is something of an animal in him. He walks without making a sound; all his movements are noiseless. I am not in love with him but he is the first man in years to interest me apart from *him*. As I said, I am afraid, I am about to have an unpleasant gift. I must get rid of it even if I have to ask the "Reich-Heinrich" personally. Kurt is going back to Vienna.

3rd July, 1940

Went for a drive with Kurt.

4th July, 1940

Went for a drive with Kurt. We mustn't do it again. Of course, Th. saw us and he is the biggest gossip in town.

4th July, 1940, evening

Got a short letter from him which made me quite crazy. *He* is really the only one who knows me. I read it again and again: " You are terribly sensual and yet control yourself—you are full of soul as no one else and yet made of fire: seduction personified." He wrote to me late at night as always and this time " at a card-table over which decisions of world-wide importance were made." I shivered when I read the letter for the first time. Kurt must go and *I* must go to Berlin.

Berlin, July 5th, 1940

Bauer (*Hitler's pilot*) flew me here yesterday though for two days there's been a ban on flying over the capital. Just before landing he frightened me; with a dead serious face he said: " I have forgotten to notify the A.A. people, they're going to fire on us." Of course it was only a joke. Tremendous excitement in Berlin, all the people up and about. Keppler met me at Tempelhof airport. Adolf is to enter the capital to-morrow—in state. And he ordered me to come so that I can see it and admire him. Thus I got rid of Kurt without any complications. He can dream now of other women.

Berlin, July 6th, 1940

I am terribly proud of *his* telegram. " Entering the capital of the Reich, I'll think of you above all." I cannot understand it at all.

Evening

Meerwald came to fetch me. I sat between Doehle and Bohle. At first they put me with the foreign diplomats. There was general amazement at the presence of an American diplomat. Handsome Alfieri (*the Italian ambassador in Berlin*) naturally cannot greet any woman without touching a certain part of her body in his sly way. With me he certainly does it regularly and then he laughs so stupidly that I can't even take it amiss. The bald Hungarian crane Sztojay (*Hungarian ambassador in Berlin and later Prime Minister of Hungary*) wanted to find out in his operetta-German whether there were any peace negotiations between Germany and Britain. How should I know? K. saved me in time from this dangerous company and put me between two reliable companions who neither attack nor question you. The ceremonial entry I'll never forget. The terrific impression is still with me. Pity there was no Wagnerian music instead of the military bands, but perhaps it was better so.

. . . Leni is awfully self-important and those who weren't in the know might have thought she was the most important person. She was surrounded by about thirty men with cameras. All were dressed queerly, as if they had broken loose from the studio. I hate her. She can do nothing but wag her four-letters. But you can become famous by it. I'd give anything to know whether it's true that she danced nude in the Berghof. But of course no one talks to me about her. A. has a great weakness for her and I wouldn't advise anybody to abuse her to his face, or even joke about her, which he can bear even less. Even Hermann, who enjoys the privileges of a court jester and permits himself more than anybody else, was silenced with a murderous look when he called her " the crevasse of the Reich "—which is a general nickname for her. To me she is slimily friendly but perhaps I am merely a matter of indifference for her. For her the main thing is that people should *believe* she has an affair with him. It is true that in private matters he listens to her more than it's good for him and she has a bad influence on his decisions on so-called cultural questions. Thank God, he laughs at her whenever she talks of politics. If it were not so, it would be the last straw!

In June Hitler wrote to Eva that the war was approaching its end. What was his opinion in July, after the French campaign had been brought to a successful conclusion, when he stood at the height of his military power and had entered the festively decorated capital in triumph? Perhaps, blinded by success, he had *acquired a taste* for more of "his" war?

Evening

At six o'clock I went to the Reich Chancellery, quite alone after all these weeks of tension! I had quite forgotten the existence of Kurt when I saw "his" wonderful eyes. He caressed my hands and spoke of Clausewitz. "Germany has won all the battles in all the wars—but lost the wars only too often. It was thus in the world war. The Germans do not realize that politics are more important than generalship." I was confused by the heat, the crowds, the dust, the excitement of this day; I felt feverish and wanted to know whether the war would be over soon. He withdrew his hands, looked through the window at the steamy sky and said slowly: "No, it won't be over so soon, not even in a few years. I see it losing itself in the infinite, in the desert, in Persia, in the Russian steppes . . ." I must have looked quite silly for he smiled, as only he can smile, and said: " It will be a great epic, believe me, a great epic." Then Keitel was announced and I had to go.

Berlin, July, 1940

Kurt has come after me and visited me in the hotel. I am terribly nervous for I haven't sufficient strength to send him away—but, to be quite honest, I am afraid I simply must have a man . . . a real man, at last. Kurt is so much of an animal that he can only cause excitement in a woman, not love. But these " excitements " are the worst for me. In any case I am starved, and again and again whipped into tension. I told him he must not come to the hotel. For now I feel that even I am being watched. No one can tell about these swine. Everyone has his own secret police, Goering, Goebbels, Ribbentrop, quite apart from Himmler who is entitled to *his*.

Monday, July, 1940

Kurt accosted me in the Tiergarten. He said he wanted to paint me. Is that all? I asked. " Yes, but in the nude," he says, the amazing fellow, and laughs with his white teeth. I told him: " If you make such a bad joke again . . ." And he replied in his singing Viennese voice: " Joke? I really want to paint you. Your breasts . . . they are unique." He might be right but how does he know? " Any man has eyes and a nose," he said and laughed again. His laughter really excited me, his lips, his . . . " And then," he went on, " you can feel it, too, if you walk arm-in-arm. With you . . . it's really lovely." I was speechless. But I couldn't let him be so familiar, all at once. He said " *thou* " to me all the time. I simply ran away.

Joachim von Ribbentrop began his career as a champagne-salesman and then married a Henckell, daughter of the famous champagne-firm. He can be best characterised by a little story. Bettina, Ribbentrop's daughter, was to meet her mother in an hotel after an absence of seven months. When she wanted to fall on her mother's neck with the cry: " Hallo, Mummy!" Frau Ribbentrop held her away from her and said seriously: " Child, have you forgotten your education? Our greeting is 'Heil Hitler!' " That was the Ribbentrop family.

Thank God, the exciting days are over. I was terribly nervous. And I had every reason to be. I really don't know now whether *he* suspects anything or even knows or whether it is only an accident. Already at the reception Ribbentrop gave, he behaved strangely. I only saw him once since his return from France and met him in full public view where he greeted me with especial kindness which he doesn't usually do and which made me puzzled. But perhaps I am just seeing ghosts. There were several hundred people at the reception. The garden is beautiful but couldn't be used because of the black-out. Joachim was stiff as always, imitated Adolf, especially in his greetings and treated the Balkan displomats with such superior rudeness as I would never have thought possible. The Rumanian attaché is a handsome black boy who made a shy attempt of flirting with me. But I sent him packing; I have really other worries. I did not want to speak to any man—not to awaken his ever-ready suspicions. I had the feeling that I had put my head into a noose after all that has happened. I was so nervous, anyhow, that I couldn't have given any sane answers. Ribbentrop exasperated me terribly. He was turning somersaults in front of the Fuehrer: "Your German style, my Fuehrer," he said once when they were discussing Adolf's speeches, "will last just as long as Shakespeare's English or Racine's French." The Reich Press Chief nodded violently and only a general who was standing by was silent; he looked so cool and stiff that for a moment I was frightened. *He* of course noticed it, too, as he notices everything and turned brusquely and in annoyance to other people. At one o'clock he left the house. Five minutes later the warning went, but I left anyhow though everyone had become a little hysterical. Naturally, nothing happened. I waited the whole night. He did not send any message. About six o'clock I fell asleep in my clothes . . .

Munich, August, 1940

I must get up to-day for "he" sent me word he was coming this evening. I look like a scarecrow. These last days have taken a lot out of me. I must go to the hairdresser. I'll never forget the week since R.'s reception—even if I live to be a hundred. What a storm of passion! I ask myself: am I worth it? I am really the "secret queen" and now I believe myself safe from all earthly temptations. "Like Dante," he told me, "you have passed through hell." When I left Berlin nine days ago, I was twenty years younger inside. Just before nine, Bauer came to the hotel to take me to Tempelhof where *he* arrived soon afterwards. We flew to Berchtesgaden. During the flight he was unusually silent.

His plane had special insulation for he liked to talk during his flights. "It's in the air I think of the most important strategic solutions," he told me once and added that he ought to take part in some air attacks on Britain in order to solve the problem of the island's conquest. But of course that could not be done. I made several atetmpts to rouse him from his mood but in vain. Once he looked at me searchingly and asked: " Have you made any interesting acquaintances?" Throughout the flight I was tortured by this terrible question which might have been meant quite harmlessly. For God's sake, why did I get involved in this horrible story with Kurt? What does it matter, I told myself, if he draws me half in the nude and then gives me the picture? Of course I should have known what would happen. I should have known that a half-undressed, starved woman and a handsome, fresh boy wouldn't be content with innocent dalliance. It's lucky that Kurt didn't do the drawing. I can't deny that we know each other. But no one can prove that our meeting hasn't remained—platonic. During the last ten minutes of the flight Adolf became quite lively again, as always when he approaches the Berghof. He only feels really happy there—he can be even gay. It was the same this time. He had only invited a few people whom he really liked, among them Frau Trost (*widow of the architect of the Haus der deutschen Kunst in Munich*) who has a calming influence on him; two Viennese actresses and a young Hungarian aristocrat, Herr van Ploery. I wore a low-cut evening dress which I knew he liked. The dinner was, as usual, modest. It was served very quickly—and there was little to eat. But, as usual, knowing these jokes and having a big appetite, I had eaten something in advance in Frau Dohna's kitchen. Afterwards it was very nice. It had turned cool so that they had to light a fire in the fireplace. Herr v. Ploery spoke about the Arrow Cross people in Hungary but as he soon noticed that the Fuehrer was not very much interested in this offshoot of the Swastika, he changed the subject to the night life of Budapest. It happens very seldom that *he* permits anyone to talk as long as this young man. Frau Trost asked: " My Fuehrer, what are you going to do after the successfully ended war?" He looked at her and did not reply as it sometimes happens with him so that you could not tell whether he had heard her or not. Frau T. went on: " Many people are going to write their memoirs after this war, there'll be plenty of interesting stuff to read." Adolf laughed with that biting irony which was sometimes a characteristic of his. " I, too, am going to write my memoirs and publish them." " Under what title?" asked the shrewd lady to her misfortune, for he replied: " My collected broken promises."

For a moment all were silent and embarrassed for no one knew whether they dared to laugh. Finally the two actresses grinned broadly and we others smiled a bit uneasily. I don't know myself what he meant by this. We broke up early. I was still deeply upset and deadly tired after the sleepless night, the flight and the day in the Berghof we had spent without talking about things, and without any of his caresses. At eleven we parted, and much in contrast to his usual custom, the Fuehrer remained alone. Not even Sch. was with him. I assumed that he would come up to me. I lay at the open window on the bed and waited. The night was light and cool, a real mountain night as I have lived through so often up here. I must have gone to sleep in the dark. As I was strictly forbidden to go to him, I had to wait until he appeared. I woke again in the darkness and was terribly frightened as he stood in front of me—only a shadow and yet unmistakeably *he*; I recognised him by the way my heart started to beat. I wanted to get up but he only said: " Stay!" and his voice was hoarse. The windows were now closed and outside the wind was whistling. I felt awful, my limbs shook with nervousness. When Adolf at last started to drum with his fingers on the table, it was a great relief. Finally the long-awaited question came, and yet I wasn't prepared for it. He merely asked, "Who's this Viennese?" But he stamped with his feet, and turned with a wild jerk so that the chair fell with a crash. In a single stride he was at the bed and I thought he was going to beat me which wouldn't have been so bad, perhaps; but he didn't do it and it was in vain that both my body and my inside winced. He only said: "Speak!" His power over human beings is really uncanny. I told him everything about Kurt except what had happened in the last few days and I would have told him that, too; but I don't think he wanted to hear any more and in any case, he probably knew it. He was quite still but so tense and watchful that I felt quite paralysed. Like all his decisions, this, too was taken in a moment. I knew that something terrible was going to happen. He stood in front of me and said, without touching me: "Of course, you're coming along." I got up, feeling weak and asked only: "What shall I put on?" He did not reply, tore open the door and hurried outside and I followed him, almost at a run. The wind had risen to a storm. Five minutes later the car roared away, direction Munich. I was still wearing my low-cut evening dress; he was in uniform without a cap. I felt so cold that my teeth began to chatter. As we reached Munich, dawn was breaking, a wonderful stormy day. We stopped in front of Kurt's flat in the Defreggerstrasse. Adolf jumped from the car, without paying any attention to me and without closing the door. Two SS. Storm-Troop Leaders stood in front of the door in long leather coats. They recognised the Fuehrer, saluted stiffly and one of them opened the front-door. He rushed inside, I always behind him, up two floors; then he rang at the door, again and again, until some dragging steps came, and someone asked in a frightened voice what he wanted; *he* roared back in a voice which brooked no resistance and was not used to any. The door opened. What followed was terrible, terrible, terrible. Kurt had hardly time to sit up in his bed, his hair tousled, blinking in the garish light; before I realised what was happening, it had happened. Down below the two SS. officers still waited. Adolf hissed a few commands to them which I did not understand in my confusion, then we entered the car and roared at a hundred and twenty kilometres an hour back to Obersalzberg. On the way, the sun rose — and it was really blood-red. I had either fainted or fallen asleep.

Munich, Wednesday, August, 1940

I was so tired in the morning, I could not go on. I went to the hair-dresser and the masseuse; I had my face refreshed and I look half-human again. The furrows around my mouth become deeper and deeper. As long as *he* calls them dimples, it does not matter. I can't say that Kurt's death meant anything to me spiritually. I did not feel any pain, only that there is something dark and terrible in the sudden ending of a life with which one has been connected physically, for however brief a time. Kurt has taught me a horrible lesson. Now I know once and for all that there are no other men for me, neither inwardly nor outwardly, now or later. I have been spoiled for other men. I felt that with Kurt, too. It won't do for me . . . I am used to other things. With Kurt it was simply a matter of nerves. I believe *he* knows that, too, just as he knows everything — that's why he doesn't even blame me. Only the "object" had to be removed. He raged and raged in the Berghof but I no longer felt afraid for I knew somehow that the storm was over. He ordered every trace of Kurt's existence to be destroyed. His 67-year old father is being sent to Dachau — third-degree imprisonment — his other relations are to buried in the Jewish city of Theresienstadt and his name is to be removed from all registers with the corresponding pages. The two SS. officers are to be transferred to a shock-troop and must not return; what's happening to the people living in that house in the Defreggerstrasse, only *he* knows. I believe he has forgiven *me*. In any case he does not talk of the whole affair. With Kurt's complete "removal" apparently everything is finished. His jealousy is peculiar and superhuman —just like everything about him. I feel that he is terribly, immeasurably jealous but he does not want to admit it because he thinks that he has grown far beyond any physical jealousy in his greatness.

Wednesday, night

He was here and only caressed me. Again and again he says how he loves the chamois leather underwear. "No one can imagine," he says, "how wonderful it is to feel this peach-like softness and your skin at the same." Can any other man in the world say such pretty things? All men are common and vulgar. I believe that what I wrote in the afternoon about his jealousy is absolutely right. In the evening he told me: "You must go to the doctor to-morrow." Of course I understood that he didn't want to come close to me until he was certain that nothing had happened with Kurt. Or perhaps he thinks I might have some disease—or a child? But a child you can only feel later . . . But he is sometimes incredibly naive in such things. "I want you to see a good woman-doctor," he said later. I asked at first, like the silly creature I am, " why not a male doctor?" His blue eyes turned quite dark and this always means something terrible and now I realised that he did not want a man to examine me. But it only lasted a moment, then he said quietly: " Of course, a male doctor. I don't trust women and I don't know whether I won't bar them one day from the study of medicine." I know that he does not trust intellectual women. But this time there was something else behind it. He had overcome his jealousy.

When did Hitler turn away from the policy of the Russo-German non-agression pact of August, 1939 ? When did his circle begin to notice this change? Gafencu, the Rumanian Ambassador in Moscow, maintains in his " Préliminaire de la guerre à l'est," that the Fuehrer decided on the Eastern campaign only after the failure of the air offensive against Britain, *i.e.* about February—March, 1941. Here is some testimony from August, 1940.

August, Thursday afternoon

For the first time in my life visited a gynæcologist. The professor was charming and really tactful. I thought how strange it is here, quite opposite to the usual: all the parts of the body that are normally covered are bare, and the other parts are covered. Naturally everything is all right. I have learned that it is possible to discover with an injection whether a woman is pregnant or not before any examination can prove it. Immediately afterwards I went to *him.* He was terribly sad, I felt it though I don't know why. He didn't ask me about the professor though I told him at once that I had come from him. Why did he send me there then? Contrary to his habit he talked little. He only caressed my hands and my hair and said he was going east in twenty minutes—to stay there for a long time. I felt quite weak. "What shall I do while you are away?" I asked. He said "Be good." And that was my goodbye. The "be good" was clear enough. I know now, too, how I am being watched. I am quite a danger for those who watch me; for if they report anything unpleasant, they are sent either to concentration camp or to a suicide-company. I went to the Braunes Haus where Sch. told me the Fuehrer was dissatisfied with Russia and had to study the situation in the east. It is not impossible that a war might start there. Sch. is very reliable, that's why it shook me up so much. War against Russia? That's unthinkable. I am going to Kitzbühel tomorrow.

Summer, 1940, Sunday

For months there was a thick text-book of surgery on his desk and I noticed that he had read it again and again. I couldn't imagine what it meant. Every time I or someone else approached, he closed the book immediately. But yesterday it remained open—probably he forgot. Now I understand everything. I read the chapter about this funny phimosis—of course, secretly. The surgeon says: operate. I say the same thing. After all, this is not a normal thing and I have read about it in the case of Louis XVI. After he had a small operation, everything became quite easy. With him, I am sure, it would be the same and would have the same effect. But he is very embarrassed and it will need a lot before he can be persuaded. In any case, I am glad that he has given so much time and study to this problem.

During the discussions in September, 1940, Hitler had certainly decided to "liquidate" the Russo-German non-aggression pact. The guarantee of Rumania's frontiers was a deliberate, open insult to Moscow. The following notes show how much and how far Hitler counted on the armed reaction of the Russians. When Molotov sent for von der Schulenburg, the German Ambassador and asked him against whom this guarantee was directed, Schulenburg declared with a smile: "Of course, not against the Soviet Union—for the USSR, we presume, has no aggressive intentions or territorial demands on Rumania." Molotov, naturally, had to assure the German Ambassador that this assumption was correct.

28th September, 1940

Discussions the whole day long with Ciano who doesn't quite know where to put me. I think he takes me for some relative of Adolf and treats me accordingly. I see clearly that as a woman I am simply non-existent for him. Never before have I seen a man who has so much charm and such bad manners. Someone has succeeded in taking a photo of him putting his knife into his mouth and this picture is being circulated. Adolf told me: "Ciano is a scamp. I don't like this type. They are unreliable. I like his wife better. She is of good stock. If she weren't, I would have dropped Ciano long ago in spite of the Duce." Ciano makes no secret of his dislike of Germans. After lunch he demonstrated on the terrace what sort of parade-step he would introduce in the German army and then told a vulgar joke about the Axis. Joachim became icier every moment and in the end he left Ciano standing, after greeting him with "Heil Hitler." At four, the two of them flew to Vienna where a conference on the Balkans is being held. Adolf rubbed his hands when they left and said: "You don't understand anything of it, of course. But listen—to-morrow there begins a new war for us, perhaps this will be the real war . . . with the Soviet. If Moscow swallows the guarantee of the Rumanian frontiers, we'll have an armed clash with the Russians within a year." "And if it doesn't?" I asked. "Then we'll start the day after to-morrow. But they'll swallow it, they'll swallow it. In any case we must be on guard." He turned round and roared, masterful as only he can be, but quite gaily: "Keitel, come here, we must talk about the east. I am no longer interested in England. That's Goering's business. Come on, come on." Afterwards we had a quiet evening alone, the others ate in the dining-room, and we had some peace. He went on talking and talking, as always, burning with some inner fire and full of restlessness. All these talks were fundamentally soliloquies. He said that no one understood the war except him; no one knew how careful Germany must be not to let victory slip through her fingers. "If England capitulates this winter, the world is ours, we'll beat the Russians in the spring. But if she doesn't give in then the Russians will try to beat us." He went on to say that he had prepared the war against Britain carefully but a landing was impossible, only Goering could think up such mad ideas. "What's the use of the Luftwaffe, the nerve war and Himmler's saboteurs if they can't put Britain within the next few months in the same condition as Holland, Belgium and France were when our troops attacked?

Perhaps it will be possible to liquidate Churchill with some of his associates; Himmler has already made the plans. But Himmler has the imagination of a policeman, not that of a politician. Guard yourself against him, he is immeasurably curious and that suits me for the tasks I am giving him. He is not strong, so he'll never become great, but he is the most tenacious man I know and therefore he will perhaps survive me." He went on discussing Himmler, saying what a pity it was he wasn't a general, a strategist, for that would help to end the war quickly. Suddenly he got up. Thinking of Himmler was not very pleasant to him. I don't know whether he sees through H. Someone once told him H. was the most unscrupulous, most amoral man in the world. A. replied violently at the time that it was nonsense, the opposite was true and H. was nothing but a mixture of middle-class mentality and the deepest depravity. That certainly shows no liking for the man. At eleven we looked at a film captured in Paris, twice running, it lasted until well past one o'clock. It was quite amusing — but not really. *He* was terribly amused, laughed, kept on pinching my thigh and said again and again: "Isn't it wonderful? I don't like the French, but such things they do magnificently, the cuckolded husband ought to be added to France's coat-of-arms. Nothing's more characteristic of France. In any case France has played its part as a cultured nation. The new France will be a wonderful country for tourists, just like Switzerland—and that's sufficient for people who have followed such a miserably stupid policy as the Parisians . . ." He went on explaining why French foreign policy was so stupid; I was terribly tired but he still did not want to go to bed. About four a.m. we went upstairs but there was no question of sleep.

On October 26th, 1940, two days after Hitler's meeting with Pétain in Montoire, the Völkischer Beobachter wrote : " A new phase has begun in Franco-German relations. The French people we are sure, must enter the ranks of the European states as a great power in full equality, occupying the position to which its glorious past, its great culture and history entitle France." Adolf Hitler's opinion was rather different.

October, 1940

After Montoire and Florence, A. came back for two days to the Obersalzberg. He is feverishly excited and still rages against Ciano who is supposed to have been responsible for the Greek war. " The Duce cried," he told me, " and that moved me. I told him that Ciano is going to end on the scaffold; he is a ne'er-do-well and as such dangerous, terribly dangerous in politics and especially in times of great and difficult wars. The Duce declared we were all gamblers but that's wrong, I am not a gambler, I am an organiser. The Greek war is unnecessary, it might cause the great turn of the tide. Now we'll see the strength of England and the striking power of Italy." He added that he loved the Greeks; he would achieve for the whole of Europe what Pericles achieved for Athens. He also said that he was most dissatisfied with France. " A nation of pimps," he said, " and Pétain is an old woman; I never know whether he's able to understand what I say. But it doesn't matter. I have no place for them in my new world-order." The Greek war worries him more than he admits; he returns to it again and again. In the afternoon he talked for hours on the phone with the generals in Berlin; he seems to have decided to settle the whole problem of the Balkans.

October, 1940, night

It's three a.m. *He* is downstairs, there are some girls with them; of course, they've sent me upstairs again. I have been waiting since midnight; I am sure he is coming. I suspected it from the conversation during the evening. The three Viennese girls wanted to dance. He behaves to these young girls like a lenient father. He caresses their hands and asks them why they do not get married. In the summer a little Hamburg girl was stupid enough to reply that she hadn't found the right man yet, whereupon every day a different SS. officer appeared as her suitor. This went on for 26 days until, in sheer desperation, she got engaged. Of course I danced, too; he doesn't want me to set myself apart. He naturally doesn't dance and so he is jealous. He always watches me and so he noticed that I was passionately fond of dancing. Afterwards he questioned me; he wanted to know exactly what I feel while dancing. But that's difficult to explain. He says that I am excited by it. I don't know, probably it's more the music, not the partner as he thinks. After all, one prefers a good dancer who is ugly to a handsome one who can't dance. "You'd like best a handsome man who is a wonderful dancer." I can't deny that. He sticks stubbornly to the subject. "When your bodies touch—you *must* feel something." A man apparently doesn't understand that it's different. A. went on asking questions and then he concluded: "The difference between the erotic sensations of a man and of a woman is elementary. I know by experience that a man touching the breasts or the legs of a woman, becomes excited even if he is quite indifferent to her, even if he dislikes her. But a woman must love a man to be excited by his touch. Is it true?" I said, yes. I am always excited in his vicinity.

But when I dance with Arthur I only feel the pleasant sensation of rhythmic movement. These are totally different things. We discussed this problem in detail; I believe that there is no other being in the world with whom he could talk about such things; not even with Leni or similar women. Perhaps, I asked myself once, he talked to Hess about them. H. is the only one who is allowed to tell a dirty joke to him. My night-gown is lovely, the finest French silk, thin as a breath of air; of course, with a mauve ribbon. It's almost four; I can hear him coming.

According to the majority of the reports, Adolf Hitler and *his* generals never hit it off well; this situation improved only when those officers who did not submit unconditionally to his tyranny had been removed. Eva Braun has described the way he summoned Keitel; the following passage provides another proof of the various insults, large and small which the German General Staff and its members had to swallow. Perhaps one of the main reasons for the German tragedy was the fact that up to July, 1944, no general had the courage to revenge himself for Hitler's insults. This is the more amazing as generals in Germany were educated from childhood to face death fearlessly. Professor A. von Salis emphasised once this most amazing psychological phenomenon: the combination of the incredible defiance of death on the battlefield with the complete lack of civilian courage in most Germans. Raeder mentioned the insult described in the diary at the Nuremberg Trials. Jodl also said: "He made the life of his collaborators sheer hell." General Jodl, who sometimes dared to contradict Hitler, had to endure "humiliating reproaches." On one occasion there was a "terrible scene" after which Hitler refused to shake hands with Jodl for months.

Winter, 1940

The lunch at T. was disturbed by a painful scene. But I knew that *he* could not stand Raeder. I drove there with Adolf. There were only six people. Before the meal was served, A. got up brusquely and said, to Erna's horror, without a word of excuse: "I must return to Munich immediately." He took his leave quickly. I stayed because I was invited for a few days and only R. followed him. He had hardly time to collect his gloves and cap. After two minutes Raeder returned, with his face dark red, threw his things furiously on the floor and shouted: "I am going to resign! This has never happened to a Commander-in-Chief, never to a German admiral!" He looked like a corpse and was literaly trembling. Erna, who had already recovered from the shock of *his* quick departure, thanks to her lucky temperament, asked: "Has anything happened?" "Of course. The Fuehrer commanded me to attend here to-day by special order after I had tried for nine days in vain to discuss an urgent service matter with him. Yesterday he told me on the phone, he only had 45 minutes for me, and I should meet him here. When he left, I ran after him and asked him where we could talk without being disturbed. He looked at me as if I were mad and said in an icy voice—in front of the guards: 'Do you think I have time now to listen to your silly prattle ?' "

L., 1940

Visiting Gauleiter M. (*Mutschmann*).

A wonderful house with old furniture and lovely

carpets. Only the pictures all seem to date from the beginning of the century. Frau M. has a handsome fair-haired Finn as a lover—and she makes no bones about it. He sits beside her at the table, and plays master of the house; he is served first and with the best. The Herr Gauleiter looks sour but resigned and waits patiently until his turn comes. The Finn behaves with perfect good manners. He is supposed to be a master in the newest love-making and Frau M. is evidently his devoted slave. Dr. Z. says the Finn takes very thin needles, fastens a silk scarf upon Frau M.'s bosom and tears it off with a sudden jerk. This is part of his "love-play." *I* wouldn't be so excited about that. M. consoles himself with his secretaries. He has an excellent collection of pretty girls—I saw them myself yesterday when he took me to the hotel, talking in his broad Saxon dialect. An Italian woman-journalist told me in Berlin, that M. used his left hand to caress his secretary while he signed letters with his right. This has become so much a habit that he had tried the same thing with her when she took him the copy of an interview and stood beside him while he read it. It must have been very amusing!

Berlin, 1940

A small party in Frau G.'s honour at the Reich Chancellery. She is considered generally as a great artist; I don't understand why, I can't share this opinion. I always find her cold. I was glad that Adolf is of the same opinion and told it to her face. Of course, she went quite pale when he said: "You'll never be a really great actress. You have too much intelligence." She refused to accept this and for the first time in my life I saw a human being protesting energetically against being called too intelligent. Anyway, she said, she could not imagine art without intelligence. But he dealt with her in his superior way: " This isn't a question of art, but of acting," he said, " and in my eyes acting is no art. Art can only be the truly creative, never the imitative." " And what about music, don't you consider Furtwaengler an artist?" asked Gruendgens. " In music it's different. The world of sound is not as firmly fixed as that of words. Sound must be freshly created every time and it cannot be produced by anyone like words . . ." " Do you think, my Fuehrer, that Paula Wessely or Greta Garbo have greater passion than I?" asked Frau G. " No, but they think less and they both possess something which isn't your gift: the music of the voice . . ." That was the end of it and Frau Goering kept silent. *He* started to talk about other things.

1941

At the beginning of 1941 there was only a single narrow front-line, along the Greek-Albanian frontier. In North Africa fighting was of local importance. Of the mighty anti-German coalition of 1939, only Great Britain remained, hammered by the Luftwaffe—but refusing to fulfil Hitler's hopes by collapsing. In Belgrade and Athens, German aggression found armed resistance, showing that the small nations still believed in Britain's final victory. On June 23rd and December 9th the war became truly global by the entry of the USSR and the United States respectively into the fray. Germany's real death struggle began. True, it started with triumphant advances in the east and south-east. In this year, as before, Eva Braun mainly records personal experiences. World politics and world wars pass her by. Even Adolf Hitler, on the eve of the outbreak of hostilities with the Soviet Union, still finds time to inspect his mistress, at length, lying on her stomach in the sun, and to take peculiar photographs which are to console him during his absence from her.

Berlin, 1941

Met F. at Professor K.'s. He has written the book " Ethics in the National Socialist state," which has been officially recognised by the Party. A shrewd, witty old gentleman who apparently leads an extremely active life for someone asked him : " How is it, Professor, that you do not at all observe the moral rules you have formulated?" The old man grinned like a sly devil, blinked with his small, shrewd eyes and said: " Have you ever met a sign-post on the crossroads that was prepared to walk with you on your way? Isn't it quite enough that it shows you the right road?" Not so bad. I also met the first man who was decorated with the Knights' Cross with swords and diamonds. In three hours he did not say a word. A slim, brown-faced man with greying hair, all will-power. I told K. that this " stone guest " made me nervous but he replied: why should a standard talk?

Obersalzberg, July, 1941

I lay on my tummy in the sun up on the upper balcony and pretended to be asleep. At first he looked at me for a long time, quite still, as I lay there naked; then he sat down by my side and caressed me, very gently and softly. He loves to do this—and always backwards. He took a photograph—back view. He says, of course, that it's only so that no one should recognise my nude photo if it should happen to fall into wrong hands which might be possible. But I know better! He is always the same. And I know why he needs such photographs. I don't know whether he suspects that I see through him. In many things he is far more naive than I, just like a child — and then he uses such funny excuses! At first I felt an inner reluctance to be photographed in this position but then I thought it was better that he should have such a picture of me when he was away than that he should take himself another girl-friend.

12th August, 1941

The evening in Karinhall (*Goering's estate*), was really nice. Gruendgens was trying to flirt with me rather violently. What does he really want? I am not a young man! I don't know why almost all men tell me dirty stories. G. is the sort of chap who could go on for the whole evening. He told me at least fifteen jokes about Emmy who is a stiff conceited goose. Gruendgens says that she hasn't the faintest idea how to spell Shakespeare and she only discovered that there was an *oe* in Goethe's name when she saw how they spelled Goebbels. G. also said that Emmy's child was fathered by the aide-de-camp. Goering knew of it, too—what he did not know was *which* particular aide-de-camp. It's certainly not Milch for he has Jewish blood and the only genuine thing about Emmy is her strong dislike of Jews. She says she gets cold shivers if she has to shake hands with a Jew. I believe that. I watched Adolf in Bayreuth, when he greeted the Jewish wife of the great singer Lorenz and kissed her hand. He did it really like a *grand seigneur*. He likes Lorenz because he is said to be the best interpreter of Wagner. He even wanted to take singing lessons from Lorenz. But the war stopped all this. Emmy, of course, also greeted Frau L. But it cost her a great effort. Hermann, by the way, was really in brilliant form yesterday. I was very pleased by one thing, because Emmy is such an awful fusspot. Hermann was just describing the bombing of an important bridge in the East by his Stukas. Goering, drinking too greedily, spilled a few drops of red wine on the really beautiful damask table-cloth. Emmy gave him a furious look. Hermann noticed this, was annoyed and did the following: He was just describing the course of the river. He took a bottle of red wine standing behind him, from its cradle and poured a huge pool on the table-cloth. Then he dipped his finger into it, drew the course of the river and said: "Here was the river and here the bridge. So the planes had to attack from this side." And so on. He drew everything with red wine on the damask table-cloth. I liked it—and so did all the others. Emmy was quite green in the face — perhaps elsewhere, too. Only Gruendgens grinned and said, Goering had stolen this scene from Bismarck. A pity. Goering likes to imitate Bismarck in many things. I don't know why. On the staircase Dietrich, who always has to make himself important, whispered to me that news had come from Hess and his liberation was near.

November, 1941

I read that Schiller had to keep rotten apples in his drawer while he was writing poetry and Gerhard Hauptmann can only write by candlelight. *He* almost always writes after midnight; the dog, this ghastly animal I hate, at his feet, in front of him the lamp and the clock at which he looks constantly and nervously as if he were afraid of losing time.

He writes very evenly, without corrections and when he finishes a page, he throws it on the ground where it can be collected and put in order later. He has the loudspeaker going all the time and it's always the music of Wagner, Wagner, Wagner, of course turned down quite low; you mustn't hear even the clicking of the automatic record-changer or he'll start to rage. I think he has the best record-changer in the world and he uses it a lot. He can listen to two Wagner operas straight off, from ten in the evening till eight o'clock in the morning and he writes all the time: speeches, orders, drafts of laws, queries addressed to Himmler, Goebbels or to the General Staff. At eight o'clock the loose pages are put in order and then they go through the teletype machine while he sleeps an hour or so. In the meantime Hinnes cleans the dozen pairs of shoes which must be ready every day, shining beautifully. When he gets up, he has a shoe parade. He walks up and down in front of them a few times and looks at them as if they were living things. With his over-sensitive, always damp, blistered feet he feels bad unless he wears the most comfortable shoes possible. Then he has a certain system of using some apparatus against flat feet which no one else understands. For a week he wears stronger and stronger supports every day and then weaker and weaker ones for another week. He thinks that this helps; but in between there are days when he wears no supports at all; he has invented all this himself and no one can do it right for him.

Autumn, 1941, Sunday evening

It is quite impossible to discuss his parents with him. He must have some kind of complex. I have tried several times—until he became angry. It took a long time before I understood. He would prefer it if he hadn't had any parents at all! He would like to be a homunculus, a test-tube creature, coming from the mysterious nothing and returning to the mysterious nothing, as Esser said.

1941, Saturday

The whole day long he was bad-tempered and angry because he broke one of his jacket-crowns at breakfast, This eternal trouble with his bad teeth really gets on my nerves. If there's something wrong with his teeth, his whole circle, the world and the war feel the result — but most of all, I. Only in the evening when he had asked me, as usual, whether I wasn't " too hot in my clothes " did I feel that his bad mood had passed. He didn't even take off his boots; he had to go on working later, he said. But then he did not do any work after all, and instead we talked for a long time. " I have been pondering since 1933," he said, " whether I should have, whether I could have, whether I must have a child. The children of great men have always been failures in history; but history might prove the rule with this one exception. In so many ways I am outside the usual historical conception that I imagine the possibility of begetting a son who might continue my work as my successor. I know that I am going to live long, Germany needs me; thus, I would have time to educate my son. Perhaps it is my duty to give such a son to the Reich." I felt quite dizzy at the idea that *I*, Eva Braun, of all people might become the mother of such a son. I really did not have the courage to put any questions but I started to tremble. *He* noticed it, smiled and caressed me and said: " For years I have waited for my doctors to find some way of determining the sex of a child.

I can't permit myself to produce a girl. A little she-Hitler—that would be ridiculous." I realized that nothing would come out of this. He only does things he simply has to do, all his decisions are made with lightning speed. Where he hesitates, little can be expected, both in great and small things. I could only try to trick him, keep it a secret—and then tell him if it's a boy. But I haven't the courage. "The idea of creating a dynasty of Hitlers has something fascinating for me," he said. "I don't want to make them Kaisers. The Kaiser-idea is too degenerate, too out-dated and dusty. But one could make them princes, that sounds well: Prince of the Great German Reich. My son could have a son again and in this way there would be Hitlers who might see the full development of the Third Reich's glory. Of course there is only one thing against it—I am a unique phenomenon which cannot be repeated; my successors could only be imitations, copies. And a copy is the more ridiculous the closer it resembles the original. A 'little one' bearing my name would be a bad joke in world history." I dressed again and we went to his study, he wanted to hear "Tristan." As soon as we left, the valet rushed in and made up the bed. That is, of course, a strict order of *his*. The bed must always be tidy, he hates crushed pillows and cushions. But I cannot get rid of the suspicion that Hinnes stands for hours outside the door, listening and that's terribly painful. *He* of course never thinks of such possibilities and *I* don't want to tell him. But how otherwise could this fellow come into the bedroom a minute after we have left it?

1942

The appeal of Hitler to the German people on January 1st, 1942, contains the phrase, so cunning in its vagueness and yet promising everything:

"At the fighting fronts the hour of new action will strike to *complete what we have begun.*" At the beginning of this year, the Eastern front was at least 600 miles from Germany's frontier. The spring and summer brought the German troops new advances, the conquest of Sebastopol, the Donetz basin and enabled them to threaten the oil-fields of the Caucasus. Rommel advanced within a hundred-and-twenty milees of Alexandria. In spite of this, in the winter of 1942/43 the German people went through their greatest international crisis since 1933. This was the year in which the Germans gave up sixty-three million fur-garments for urgent collection, in which all class distinctions were removed in the armed forces, so that from then on any Nazi could become an officer without higher education. At the same time a " thorough clean-up " took place in the army and Adolf Hitler took over " in the darkest hour when one disastrous report followed the other, the personal responsibility for the fate of the German armies." For the first time since the outbreak of war, food rations were cut considerably on May 6th, 1941; whereupon Goering remarked threateningly: " If any nation must starve in Europe, it won't be Germany." But on February 24th, the official army communiqué reported: " South-west of Stalingrad and in the big Don bend the Russians have, by throwing in war material and troops regardless of cost, broken into the defensive front on the Don." This huge Russian attack could not be stopped as before. The tragedy of Stalingrad began. On October 25th, Rommel had to start his retreat in Africa. February 14th saw the relief of Tobruk, the 20th the loss of Benghazi. In the meantime, on February 8th, British and American troops had landed in great force in North Africa. According to Eva Braun's diary it was in 1942 that Speer mentioned for the first time an explosive which " is connected with the atom." Did that mean the atom-bomb as early as that? In 1942 the German people began to doubt their leaders for the first time. The four would-be assassins whom Eva Braun mentions were perhaps symptoms of an awakening resistance to the Nazis. But the Nazi propaganda succeeds once again to bamboozling the German people and Adolf Hitler can state with satisfaction: " The German people is the best disciplined in the world." For Eva Braun this year brings a different sort of sensation.

20th April, 1942

According to orders I have been waiting here on the Obersalzberg since yesterday. There is a south wind, I have a head-ache and I am afraid I won't be very lively when he comes. There are rings under my eyes, I can't get rid of the fear that I won't please him now. They'll be congratulations raining upon him in Berlin; I only sent a brief telegram; I can never think of anything to say on such occasions. Goebbels, of course, thought of something. He sent me a really wonderful pearl necklace and wrote: " The prettiest gift I can present to my honoured and beloved Fuehrer is this small token of my respect for you." The " small token " has eighty-eight perfectly matched pearls and must have cost at least a hundred thousand marks. Goebbels with all his brains, is still an impossible man; he still does not know that Adolf hates pearls and that

I am never allowed to wear them in his presence. He says they bring bad luck.

21st April, 1942

Telegram from H.Q. " Cannot come." Nothing more. What's happening? They say we are in a bad spot! Yesterday I rummaged in some drawers and found a photo of Hess. *He* often dreams of Hess. I know, because he has talked of him in his sleep, but he would never confess it. He has only mentioned him once and then in a tone which still makes me feel cold. " One of us is mad," he had said. " Either he or I, he or I . . ."

22nd April, 1942

He came after all for two hours. He did not touch me and said: " I only wanted to look into your faithful eyes." And then he spoke about the war, said that the offensive was about to begin, that Russia would lose a territory inhabited by 100 million people. " That means the collapse of Moscow." He wanted to know who had told me that we were in a bad spot in the East. But I couldn't remember. He seems quite possessed by the idea of the East. As I looked at him, I thought, this is no longer a man but a piece of white-hot steel. He has really become a stranger to me and I can hardly recall our intimacy. When he speaks of the " East," there is something so distant in him, I hardly recognise him. He wants to build, build again. At the beginning of the Russian steppes he plans to erect the " Gate of the East," a huge simple edifice, the radial centre of seventeen Autobahnen running from east to west carrying the wheat of the Ukraine and the petrol of the Caucasus to Europe. Once he told me the war might lose itself in the infinite. I feel we have got that far now. To-day no one will doubt the Fuehrer's greatness. I told him that he was certainly more important than Napoleon. He only said: "I don't like Napoleon. I really don't. He was not timeless. I only like people who are timeless like I am. Jesus was outside and above all time, but he was probably a Jew and so there is no comparison . . ."

The visit to Sonthofen was well worth the trouble. (*The "Ordensburg," Sonthofen, was the most famous educational centre for the new German élite. Only boys from the most trustworthy SS. families were admitted.*) What wonderful boys are growing up there! A look at their faces and eyes is sufficient; one knows that there the future Herrenvolk of the world was being shaped. Amazing that it was fat Ley who thought up this idea! Dr. Wiedermann was courtesy itself taking me around. "We Mauretanians," he said, "are already a chosen race, only people belong to us who have proved their mettle a hundredfold and more in mortal danger. But if *we* begin now to form a new leader-circle from the best biological material of Germany, only good results can ensue." I naturally did not know what Mauretanians were. "That's an idea set out in Ernst Jünger's book, *The Marble Cliffs*. Fundamentally we SS. Leaders are nothing but German *Politruks*." Everyone is reading this book. I don't know what's so important about it. I couldn't finish it, I found it too high-faluting. Axmann told me that Jünger had confessed to him that in the book the "Chief Forrester" was not the Fuehrer, but naturally Stalin. I found that out for myself. While half of mankind seems to discuss *Marble Cliffs*, the other half argues about Schiller's *City of Lycurgus.* I have read it, too, and it made me desperate. This story with the soup of blood and the murdered babies, etc., is too horrible. And I can't find any parallel with our age—where on earth are babies being murdered? Sonthofen, by the way, has its own cemetery which W. showed to me. I thought there were a good many graves there and W. said: "Our demands are enormous here. Only the very finest endure it—but afterwards they can fight death and devil. Do not forget, we create the truly universal man here — the like of which the world has never seen since Leibnitz. And even Leibnitz was only universal in the spiritual sense while our boys must be perfect in body and mind. The mortality in the first year is rather high—about thirty-seven per cent." I stood the trip well; I only felt sick once but no one noticed it and I bore it bravely, thinking of *him* and of the nice young kids who are being educated in Sonthofen and who have to endure far more than such small discomfort.

Spring, 1942, Saturday

Speer (*successor of Todt, Minister of Armaments and Supply*) was here and brought a piece of sensational news. They have discovered some new explosive in Königsberg which is somehow connected with the atom and can destroy whole cities. Of course, it's strictly secret; I only managed to catch fragments of the conversation—"decisive for the war," etc. The main thing would be the complete elimination of Britain. If the English cities could be really destroyed, the war must end for even America cannot fight on without Britain. Of course we'd still have the Russians on our hands but our boys can deal with them. For the first time it moved, I felt it . . . My God, it's a funny feeling.

I got up because I thought at first that something was going to happen. Now at last I really know that there's something alive there. God, I can't quite believe it, it's so overwhelming, the idea that I am to bear his child. No one, no one must know except him!

Berlin, 1942

Schellenberg (*Major-General of the Waffen SS. and after July,* 1944, *chief of counter-espionage in the German G.H.Q.*) invited me; he said he had something " sensational " to show. We drove through badly battered Berlin and then for three quarters of an hour in the direction of Frankfort, then we reached a tiny house which was crowded with old pictures and statues. One of the Rembrandts was supposed to be a family heirloom — the rest war booty. " State Secretary M. who plundered the whole of Europe for Hermann, got a few ' leavings ' for me; after the war I am moving them to a ten-room flat in Berlin; even now they have an international value of seven million marks, not counting the Rembrandt, of course." Sch. collects not only pictures and statues, but first of all, as he put it—positions. Not military positions, of course; no, he possesses the biggest collection of nude photos in the world. He told me he could get hundreds of thousands of dollars for it in America. But that isn't the reason he made the collection—he simply enjoys it as a hobby. In the Eastern territories and the concentration camps he was able, as he told me, to have photographs taken of unique and entirely new " positions and things." He even sent " special war correspondents " to do this and now he is starting on moving pictures! That would be really interesting, he said. He showed me various pictures from the file " cruelty " and " Jewish "; also, for comparison, photographs which Terboven had sent him from Norway in exchange for some Schellenberg's duplicates. It was all very interesting. Sch. wanted to know whether the Fuehrer might be interested in his unique collection. Of course I gave him an evasive answer; but perhaps he *would* be interested. I might tell him about it at the right moment. *He* of course does not want any outsider to suspect his interest in such things.

Though she very often spends much time in describing unimportant details in this particular year, so important for her, Eva Braun is amazingly reserved and concise. She is expecting a child; her references cannot be misunderstood. Apparently she visited Dresden twice to find the place where she would bear her child—under a false name. Did she have a child? When? where? In 1943 she returns once more to this problem.

Dresden, Autumn, 1942

In Dresden for the second time this summer; I like the town better than ever. I am happy to know that it's going to be here. The room has been reserved for Frau Hertha Lind. The papers are in order. Here everything is normal, the war is hardly noticed, even better than Vienna. No one, by the way, has noticed anything; it's like a miracle. Only Erna sometimes looks at me searchingly but I can look so innocent that even she and Frau M. who have the sharpest eyes and ears, have been fooled.

In his " Storia di un anno " Benito Mussolini declared that it was France and French resistance that opened the gate of the Mediterranean to

America. Hitler shares the same view and reveals the same inclination to put the blame for military disasters upon others — neutrals, allies, non-belligerents or even the conquered nations. The same fate is prepared for Italy. After September, 1943, Adolf Hitler even tried to blame Italy for the tragedy of Stalingrad.

November, 1942, Sunday

The American landing in Africa seems to be very serious, everybody is talking about it. A. told me: "The real, historical treason of the French has now started; they opened the Mediterranean to the Anglo-Saxons—and I am convinced that Vichy knew of it. Darlan has been treated too badly, but Ribbentrop doesn't understand this, he thinks he should have been kicked even harder. It's fantastic—I am surrounded by dilettantes. The Italians have a better idea of the position. Canaris (*Admiral Canaris, head of counter-espionage in the German G.H.Q.*) tells me that the Italians had full information about the Anglo-American preparations for the landings and even foretold that Algiers and Casablanca would be involved. Why, I asked the wretch, wasn't I told of this? He replied that it was passed on through Mackensen or someone else to Ribbentrop, but he only smiled. Now of course the Italians laugh at us—as far as they are able to laugh under the circumstances. If we were as cunning as they are and they were as hard as we, we couldn't lose the war."

Sunday, Autumn, 1942

Went with *Sturmbannführer* Hanschke to the Ufa cinema; the newsreel was very interesting, the Atlantic Wall, the Murmansk front. Hanschke told me that the SS. command had appointed him to organise an " SS. stud-school " for " pedigree bulls." Why does the SS. need such things? And are they going to school? Oh no, my question was very naive—it was all a most interesting scientific experiment. H. also told me that he was really a doctor but had given up his career to stay in the SS. Now he is to organise an institute in which healthy SS. men who have absolutely reliable political views are to be kept for experiments in artificial insemination. The League of German Girls, *BdM.* is set up a similar institute as a " breeding school." I didn't understand at once why all this is to be done through artificial insemination. After all, the two " schools " could be housed in the same building and the results would prove whatever they wanted to prove—after the usual time limit. But Dr. H. frowned and said, it's just the natural breeding they want to avoid here. Wrong cross-breeding has to be eliminated. Such experiments must be conducted according to strict scientific principles. And this is only possible in a totalitarian country. Hanschke said that America might be superior to us in material but we equalise that by our superior control of human beings. Then he told me some details. It is possible to say the most awful obscenities in this scientific jargon. The " breeding stud " is to commence operations in three months and he invited me to inspect it and to study their working methods. Interesting but too exciting. I'd rather visit one of the stud-farms.

Winter, 1942

Now that all the excitement has passed I can at last sit down quietly and write. Looking back the whole thing doesn't seem to be so terrible as at the time you are in the midst of it and when you think, it might go badly—or has perhaps already gone badly. For a moment it really looked like the "wrath of God." Adolf had discussed the position of the Catholic church in Munich for many hours. Rosenberg (*Reich Leader and Chief of the Party Cultural Office, later Reich Minister for the Eastern Provinces*) wanted to have all Catholic bishops put into prison, they were anyhow openly declared enemies of Germany; it should also be considered whether the Pope should not be restricted somewhat in his freedom. A. became terribly impatient as always when his collaborators do not understand him and roared at R.: "I know, you want to become Pope yourself; but wouldn't it be better if you took care of the Eastern territories where economically the situation gets worse and worse?" Rosenberg went pale and never said a word. "Lock up the bishops!" continued A. in his rage, "that would be just like you and Hofer, that eunuch (*Hofer was the Gauleiter of the Tyrol, notorious for his fight against the churches and monasteries; he banned church bells in his Gau*) who has so little sense that he had *my* personal coat-of-arms, my very own, tattooed on his belly! He was probably drunk. It's getting so bad that I'll have to introduce a ban on drinking for all Gauleiters—or they'll all die from seeing pink elephants. You make your politics as if we had already lost the war!" He went on accusing Ribbentrop of ignoring world opinion—which was all right when we had won the war but not before then. "Goering wants to bomb the Vatican," he shouted, "you want to arrest the Pope. Am I completely surrounded by lunatics?" R. began to say modestly: "My Fuehrer," but he was given no chance to continue. "Don't worry about the Pope, he'll give up the ghost, even Popes do sooner or later, believe me —especially when we have won our victory. During the war I want no ideological battles but shrewd politics. From now on the Church will be Himmler's department, all bishops will be under surveillance and we'll lock up or behead the people they make use of, not their Lordships themselves." And then a flood of abuse poured out, all against God and the priests; I began to feel frightened and worried myself. At last, late in the afternoon, we set out alone for Berchtesgaden and then it happened—about twenty kilometres from Berchtesgaden. I hadn't heard the shot myself—but there was the hole in the windscreen, plain to see. Adolf stopped the car which had almost turned over, in a minute; tore the gun from his pocket, shouted at me: "Sit still" and jumped out like a tiger. At the same time he slipped off the heavy leather coat to have more freedom of movement. All this went like greased lightning. I sat there stupidly, trembling in fear and I remembered, of course, that awful evening in the Hofbräuhaus when the bomb exploded and the hall was filled with smoke, fire and flying débris. But I took hold of myself, got out and wanted to look around. I was not afraid—I had no time. Adolf came back with a small, half-bald man who wore spectacles. He had grabbed his collar; from time to time he kicked him and shouted at him but so wildly that I couldn't understand a word of it myself. The little man had a revolver in his hand, holding to it grimly. He could have used it for A. had slipped his gun back into his pocket. It was really a miserable spectacle. "Beast, wretch," roared Adolf, "are you alone?" The man did not reply—not even when he was kicked again and when A. smacked his face hard. Adolf stared at him, shook his head, took his gun and forced him to get into the front seat. I had to sit in the back and was given his gun to hold. "The fourth case this year," said A. grimly and pressed down the accelerator. We drove up, the guard jumped to attention; the little fellow was taken away after a brief explanation by Adolf. To the cellar. I know what it means. To-morrow he will have neither teeth nor fingernails left, I thought. It was even worse. Adolf shot him himself at dawn because he would not confess. *He* is convinced that there is some organisation behind it. All four would-be assassins have been caught but not one of them confessed. At noon Himmler came up but by that time even the corpse had been buried. What a thought—how quickly a human being can be disposed of! The whole story ended by Himmler declaring categorically: the Fuehrer must never again drive alone, never again. And Adolf raised no objections. It is peculiar how close he has moved to Himmler recently, how he lets H. give him what are practically orders of conduct. Himmler was given his own air force—to be used inside the country if necessary. The funniest thing—no one knows who the little man was for he had no papers on him. And as he didn't speak a word, we don't even know whether he was a German.

Munich, 1942

Colonel S. told me to-day that Himmler has installed child-brothels in Poland and the Ukraine. Girls and boys of good Polish and Ukrainian families from the ages of ten to thirteen are taken to them and are at the disposal of the SS. Thus the leading class of these countries who cannot be robbed of their position by dispossessing them, is made impossible socially and in every other way. Not even the noblest Polish princess will be able to find a husband—if it is known that she was a prostitute at the age of ten! Gradually the children of all important Polish and Ukrainian families are going to be " inoculated " in this manner. On the other hand the sterilisation of Ukrainian women is to be stopped. It is the view now that they should have as many children as possible. After this war Germany needs labour slaves. Germans are supposed to occupy only leading and key positions. Manual work is to be done by foreigners.

1943

After the military reverses of 1942 which clearly represented the turning point of the war, 1943 brought serious political crises which culminated in Italy's desertion of the Axis. What a difference between Hitler's New Year messages of 1942 and 1943! In 1942 he still dared to promise the end of the war; while a year later he was content with the statement: " We possess the *right* to *believe* in victory." The conference of the Gauleiters which Eva Braun describes was already overshadowed by the realisation that defeat and the subsequent destruction of National Socialism had become a reality. Hitler's insomnia became "frightening." Once again he started to read the " Textbook of Psychiatry." Was this a sign that he began to worry about his sanity? The terrible tension under which the German people lived owing to enormous losses at the fronts, the ever-increasing air attacks and the general hopelessness of the situation, did not prevent Hitler from following his private amusements nor Funk calling together " a company of gluttons." And Eva Braun, because Frau Goebbels disturbs her peace of mind, exclaims indignantly: " This year of 1943 makes everybody mad."

Munich, May, 1943

The Gauleiter Conference ended yesterday; immediately afterwards we drove to Garatshausen, but only for a few hours. He was at first angry and forbade me to talk English with Baldur. He likes B. well enough but does not trust him. B. is too international for his taste. B. knows that and sometimes he risks his head when, for instance, he talks English in *his* presence or when he tells me dirty stories like yesterday. He asked me impudently whether I had heard of Monsieur Sixfois in Casanova and whether I had had similar experiences in my life. I got the Casanova book and read the stories and I must say, B. was more than cheeky talking about them. In spite of this I tried to defend B. "Be objective," I wanted to start. But he made one of those gestures which brook no opposition. "I never had the intention of being objective or honest," he said and I noticed that he was bad-tempered. Desperately I searched for a different subject. At the moment almost everything is dangerous for I believe things once again look bad. He is angry with his generals who desert him treacherously or deliberately do the opposite of his commands. Then he has to brood again for nights to right the mistakes. We ate alone in the park and he was quite mad—I lay on my dress but the damp came through; only I did not feel it then, but later. On the way back he was terribly loquacious, he talked and talked. "I told them yesterday quite clearly. It's not a phrase, it's really a question of victory or utter destruction. They don't want to understand me, they think it will be different . . . But now I told them: if the German nation proves incapable of achieving victory, I'll prepare for it such an end that at least the survivors will find in it strength to build up a new national existence. Of course, this will take more than a hundred years—and it will be a Germany I'd hate." His eyes glittered and his whole face was changed.

Had lunch with Esser (*former Bavarian Minister, then Secretary of State for Tourism, &c.*). It was very pleasant and there's always something special. He gave me a long lecture about Stalingrad which was no defeat but really a victory because it was possible to relieve the whole threatened Northern front. I feel I can't listen to any more about Stalingrad. A. has talked about it for nights and nights; he said it was a symbol and had to be retaken; that the whole Russian campaign depended on it. It is a terrible story. I was always hoping for the wonder-weapon of which Speer spoke to me - -the whole thing, of course, is very, very secret, but they are working at it day and night and perhaps next winter it can be used. Southern England will be "rubbed out" first as Sp. put it so aptly. And then they'll see how it works. But I was more interested in the funny story about handsome Dino A. (*Alfieri, Italian ambassador in Berlin*). Major General Klatt surprised him in Frau Klatt's bedroom. E. says that the General first thrashed the poor chap with his riding crop, then he had to leave the house in his pyjamas—luckily he had his car in front of the house. Herr von K. is supposed to have told him that if he met him again in Berlin, he would shoot him. Thereupon A. left not only the capital but Germany in sudden flight and went to Italy. Esser believes that the Fuehrer himself telephoned to the Duce—but this is only the invention of a "political mind." Telephone the Duce because of a woman who cuckolds her husband! Himmler told me that they had conducted a sort of poll into the number of adulteries in Leipzig, Koenigsberg, Bremen and Vienna. It is proposed to introduce a very severe law with the death penalty for women who commit adultery while their husbands are at the front. In each of these four cities three thousand cases were investigated without those concerned having any idea of it. The result showed that no such law could be introduced. In Vienna the results were comparatively favourable which surprises me for I never thought Viennese women particularly moral. And now they want to bother the Fuehrer and the Duce because of Frau von Klatt! Funk (*the Minister of Economics*) also told me that the members of his Ministry volunteer in droves for front-service because they want to get rid of their wives whom they married in their humbler days and whom they no longer want.

Sunday evening, 1943

I don't know whether I have done something very silly or just moderately stupid. A. arrived unexpectedly at four o'clock and found Helga with me. She was fourteen yesterday. He has a weakness for such very young girls and I noticed Helga attracted him at once. They met more or less on the doorstep for H. was on her way to the bathroom. She blushed deeply and left at once. He asked me, of course, who she was, how old she was and where she was going. She is pretty, isn't she, said I, but still like a boy—if you saw her in the bath. He looked at me in such a funny way, his eyes darkening. I knew the symptoms well and he said hoarsely: "Let's go and look at her." I was, of course, speechless, but there's no arguing with him and in order not to appear stupid, I went ahead. I was thinking wildly what to do. I believe that Helga at fourteen is a little bitch and dangerous just because she has such a boyish figure; he likes that. But I felt I must not show that I was worried, that jealousy made me yellow with envy—or everything would go wrong. Helga stood there naked in the bath, in a provocative pose as if she had expected us—or rather, him. She took the towel slowly and turned quite shamelessly with her slim hips. She *is* a little bitch. Then I had a sudden inspiration and I still don't know whether it was stupid or clever. I told Helga: " Don't be silly, you know who is here. You ought to be glad that he looks at you." But I looked at him and noticed the pulse throbbing at his temples. Yet he said nothing, he turned round and left and Helga got out of the bath a little ashamed—but only a little—to dry herself. In the bedroom I asked him: " Did you like her?" And I knew quite well that he was thinking of her. " And if I liked her—and wanted her?" I began to feel very bad indeed and I couldn't have carried on long. But I took hold of myself once more and said: " What about it? Take her, if you want to, why not?" If it had lasted another minute I would have started to howl like a child; my face was quivering and my self-control was quite gone. He took me into his arms, smiled and said: " You are magnificent. I don't think any other woman could be like this." Suddenly I was all on top, a weight like a ton fell from my heart. I started to babble. Because I felt myself I was " magnificent," because I knew that the temptation had passed for him and that the little fourteen-year-old bitch was expecting in vain to become his mistress. " Why magnificent?" I asked, but my eyes filled with tears. " I am an artist," he said, " and beauty excites me wherever I find it." " Of course I can't make myself so pretty," I replied, " but no one could be as devoted to you as I am and no one will forgo her own pleasures to give *you* more." But that was perhaps too much. In spite of this he stayed for a long time, he was more excited than usual—and just as deadly silent as always on such occasions.

The Intelligence Service has provided full proof that Adolf Hitler married Eva Braun on the evening of April 29th, 1945. As the idea of a sentimental gesture towards his long-standing mistress can be excluded, we must still enquire into the reason for this marriage. Up to the very last moment Hitler enployed the melodramatic and theatrical methods of National Socialism. He took good care that his death should be surrounded by an impenetrable mystery. He also made good his words to Eva Braun: to prepare a terrible end for the German people. Perhaps the secret of this marriage is solved by the reference to " projecting ears," which Hitler also possessed? Perhaps Germany's Fuehrer did leave behind a pretendent for the title " Prince of the Great German Reich " ?

Berlin, 1943

This year is making everybody mad. The story of Magda Goebbels makes me very, very nervous. Yesterday I went for a walk in the Tiergarten. Suddenly she stood in front of me, looking like a ghost and in a fantastic mink coat. In my own shabby pony-coat I felt like a Cinderella. I don't think it's wise for the wife of the Minister of Propaganda to go about in such an expensive fur-coat. What would our soldiers think who meet us and who have no winter overcoats? But she is mad and I cannot help her. She looked at me with her miserable doggy eyes, pleading, begging and stroked my arm. I cannot understand the world any more! After all, this woman has borne four children; her marriage could not have been a platonic one! And now she wants to spend the night at my place. This has been going on for months. She even appeared in Munich and told me she could not find a hotel room. And all the time she talks nonsense about her love for the " Fuehrer " which *she* " sublimated by transferring it to me." The devil! She annoys me because she keeps on calling me " little one." I'd like to tell her where she gets off, Frau Goebbels! All the people are laughing at us. Goering asked me how I liked my new position as *cicisbeo* in the Goebbels family? I told her: " I am a girl, Frau Goebbels." She looked at me stupidly and said: " I know, I know." What can I tell her? Her butler was just here, it is eleven at night and the second air raid warning has just ended. The butler brought a small package with a gold vanity case in it and a pink slip of paper with the words in red ink: " For remembrance." I really don't know what to do, I am not equal to such intensity. I took out some pictures of Dixi and looked at them. He is getting sweeter and sweeter and his projecting ears are really charming. Again a warning . . I must finish and go down to the cellar.

Four o'clock. It lasted a long time; all around there are fires burning. I have become quite indifferent. I have lost my appetite for sleep, I'll play the gramophone and write *him* a letter.

Berlin, Friday, 1943

About two o'clock the copy of the Deutsche Allgemeine Zeitung arrived. There was an article in it which said the Fuehrer was proud to be an Austrian. A. immediately rang up Goebbels and declared once and for all—he wanted it to be understood that he was in no way proud of being an Austrian, on the contrary, if this question was to be raised at all, he was ashamed of it. And he deprecated such articles. The German press must understand that he was German and nothing less and that all German tribes had a full right to call him their own. G. must have been stammering a lot; I could hear him whining at the other end of the room.

Berlin, 1943

Magda's letters are so silly that I really cannot follow them. I am used to men wanting to get off with me. But that Frau Goebbels tries to start her little games in the box at the theatre and then writes a three-page letter about it; that's too much, it's burdensome, even nauseating. I don't really know how she can do such things, she's so much of a prig and a puritan otherwise. People are already gossiping about it. L. says that earlier on there was always a young girl in the Propaganda Ministry whom he and she liked equally. I am not against it because of any principles, God forbid, let everybody be happy in their own fashion, but I myself find it disgusting. If she were a pretty young girl, it might be different; but I don't like this bloodless mask. I wrote to her and told her my views quite clearly and I also hinted at a threat. If she won't leave me alone, I'll talk to Joseph. He can bring her to her senses.

Sunday

Now we have our scandal; it's really the end. I went quite innocently and unthinkingly to the reception at the *Promi.* All the people dodging front line service were there. Goebbels looks awfully ill. But otherwise he is still on the top, the only Reich Minister whom Adolf still trusts and likes. He has certainly roused himself from his lethargy of 1938. And he is no longer so crazy about women, they say. I walk up to them, he comes forward to meet me, kisses my hand, but she remains behind. At first she looks past me as if I were just air, then she suddenly pulls my letter from her bag, throws it at my feet and says so loudly that hundreds of people must have heard it, for they all looked at us: "You mustn't think that I let myself be insulted." And she moves away, glittering with diamonds. On her head a diadem, at her breast a brooch as big as a child's hand, and every finger of her right hand something dazzling. Probably she wanted to show off all this to *me*, for usually she doesn't deck herself out like a Christmas tree. G. was quite startled, grinned and said: "It's play-acting, suits my Ministry, but I'd rather have a comedy." I simply gave him the letter and came home. At midnight Magda rang me up, sobbing, and asked my forgiveness; it was only her love for the Fuehrer. I could not stand it any more. Fortunately there was an air raid warning and Frau Goebbels fled to her shelter.

Monday

The Heinrich of the Reich (*Himmler*) was here and wanted to know whether I had been insulted. He came in uniform, polished and slippery as always. He is collecting material against Goebbels and even I serve his purpose—though otherwise he hasn't any use for me. I am sure he is not interested in women. I don't find him as sinister as others do. Now he is the sharpest rival of Goebbels for Adolf's favours. A pity that the court jester has been so completely excluded from the running. I would really like to know the truth about the concentration camps. I asked H. point blank. "Come and have a look at one," he said. "Which one?" I asked. "Whichever you like." This doesn't sound so bad. Of course I could not possibly go without asking *him* first. And he would never permit it. The concentration camps are Adolf's personal idea, they did not originate with Himmler. A. insists on regular reports about them and gives quite detailed instructions for the treatment of the different groups. The concentration camp brothels were also thought up by him. Of course, I didn't tell H. anything about Magda. That would have been the last straw, to turn it into a state affair!

The defection of Italy from the Axis upset much more than Eva Braun's "plans." Hitler's reaction is typical of the foreign policy of the Third Reich. He is brooding over revenge. Yet the campaign against Italy has been prepared for a long time.

1943

It is apparently written in the stars that I should never achieve the state of real well-being and true content. The events in Italy have upset all our plans. He wanted to come at last for two days and go with me to Fürstenried where there is more peace and quiet than on the Obersalzberg. Of course the putsch in Rome has spoiled everything. Now he sits up on the Obersalzberg and broods about revenge against Victor Emmanuel. Goering offered to talk to Badoglio but A. refused it angrily. He says Italy must be left alone to act—like any criminal, she will commit a mistake and then the blow can be struck. Preparations are already being made. Keitel is said to have worked out plans in every detail for a campaign against Italy already in the spring of 1942. A. told me: "Goering disappoints me more and more. He is really nothing but a sad bull mourning his lost potency. If a political leader begins to think of his family, he must be sent home and care must be taken that he dies in his bed." Long conferences with Himmler. A special group of the SS is to be sent to Italy to discover the Duce's whereabouts. They are carrying half a million Swiss franks in gold for bribing people. " I can't understand Mussolini," Adolf said. " The whole thing looks to me as if a man were arrested and put into prison by his own wife." Then he started to abuse our Foreign Office. " All fools, fatheads and homosexuals! And they think I don't know what sort of people they are! If I had the same full information about other countries that I have about my collaborators, I should have conquered the whole world, including Guatemala, long ago!" His insomnia is now frightening and I am getting more and more exhausted for he wants to talk to me all the time, talk and talk and talk. He calls me " Maedi " again now; he often has anxiety-attacks, especially when he thinks of Hess. To-day he suddenly said: " I'd like to walk at least once unrecognised through the Prater in Vienna and get on the Riesenrad (*Giant Wheel*)." I felt frightened but I couldn't say why.

Autumn, 1943, Obersalzberg

At last I met Benito Mussolini, the great Duce. A ghost-like, bald-headed man. His eyes are sad. He speaks quite good German. You feel as if you were facing a man who has already died once and who therefore knows something about the Beyond. He is certainly not a superman like A. On the contrary: there is something enormously human about him. Even in matters of sex. He measured me with his sly eyes in a certain way. He has fallen for a Salzburg countess who belongs to Adi's constant retinue. L., who always knows everything, maintains that it was a false alarm. True, he tore the clothes from Lilly but that was more or less everything. Might be true. Poor fellow, he looked so tired and he probably felt that he has played out his part. Adolf says of him: " He is my friend but I prefer my dog. The events in Rome might cost us our victory, at least they'll postpone it. Friends who are dogged by bad luck are finished for me—I must take their fate into my hands."

Hitler's sterile thinking in foreign politics and his complete inability to understand the mentality of other nations, finds shattering expression in the following conversation. His hatred of Jews and the background of this hate are also explained in a most concise form.

Winter, 1943

To-day he came again but only to read Bleuer's " Textbook of Psychiatry." He wanted to be alone with the book. It's again the same as in 1937. For more than three hours he locked himself in, then he turned the key in the drawer and put it into his pocket. There is something uncanny about him these days; I am almost frightened by him. He is more taciturn than usual and he only looks at me, from head to toe. Something oppresses his mind; he looks so harassed. He asked me what I thought of Roosevelt's speech. I hadn't read much about it. Then he burst out: " What does the fellow really want? I don't understand him at all. For days I sit over the speeches of this cripple and try to understand his mentality, but it simply doesn't make sense. Churchill I understand well enough, that's our language, I like him somehow, a pity he isn't a German, I would make him Minister of Propaganda for he isn't as preacher-like and double-faced as Goebbels. I have to get Roosevelt assassinated, he is my chief enemy; while he lives, I have no peace; I can deal somehow with the other two, I can see deep into them . . . I could tell you at any time what they plan, what they hide, what they want to say—but I don't see clearly with Roosevelt. Perhaps because I have never been to America, a pity that it's so far, so unattainable, the paradise of the Jews. I'll fetch Stalin perhaps from his nest, the experiences with Mussolini are quite promising. I must win the war—for the idea that hundreds of thousands of Jews sprawl everywhere in comfort and laugh at me, is unbearable. I must reach every country of the world and no one must laugh at me, no one, no one." People say that he is chewing carpets when he is furious—which is nonsense. But he can rage and shout so that I get terrified.

Winter, 1943

Dinner at Funk's (*Minister of Economics*)—a real blow-out, a fantastic gorging. I have never seen such gluttons in my life. It was a gathering of voracious eaters; fantastic quantities of meat must have been consumed. It was really funny to watch fat F. stuffing himself with everything that came on the table—and doing it with a lot of noise, too. His red, hot face was distorted with greed. He cracked the lobsters with his teeth and explained the difference between tails and claws, the upper and lower halves of the lobster's body. " Hardly a handful of people in Berlin know about these things, it's a shame," he declared. Opposite to me a Frenchman was sitting who had a battery of wine-bottles behind him. "They know how to eat in this house," he declared shrilly for there was a lot of noise going on, " but nothing about drinking. People drink an awful lot in Germany—but not even men of fifty are able to tell a really good wine from a mediocre one. So I always bring my own wine with me." It all ended on the sofas and I fled before I had to watch everything—but literally everything.

1944

Only in 1944, the year of the invasion and the total air war against Germany, do Eva Braun's notes show something of the over-tense mood of the Third Reich and the gradually growing hysteria of the Nazi leaders. Only Heinrich Himmler, according to Eva Braun's testimony, remained unmoved by events. His policeman's instincts triumphed in the creation of a "supervision bureau for Reich Ministers." The Tower of Babel began to sway. On June 2nd, Rome—defended with a terrible sacrifice of blood—fell; for the Allies this represented the symbolic liberation of Italy. On June 6th the direct attack on the German-dominated Continent began; after early setbacks, the defeat of the SS shock-divisions turned it into an almost unparalleled triumphal procession, leading the Anglo-American forces to the Western frontiers of Germany. But the Fuehrer, who suffered from insomnia and hallucinations, found time to send his mistress a short story which he had written! During this year one German city after another was reduced to ruins; Rumania and Finland left the Nazi camp; Hungary attempted to do the same but without success, and in a single week the Gestapo executed 2,726 men and women in Germany alone. Within the Hitler Youth of Vienna a strong group was formed to "remove Hitler," and on July 20th German officers made an attempt on Hitler's life, the failure of which prolonged the war for another eight months.

Berlin, spring, 1944

Visited Himmler at his headquarters. Gave guarantee for Frau L. and she was released at once. " It's a painful story," said H. " Since the day before yesterday we have known that her husband was innocent—it was just a similarity of names—but we can't resurrect the dead. I thought of putting her into a concentration camp—making her disappear. But she won't do anything silly, will she now?" I gave the necessary guarantee; after all, it wasn't necessary to tell her that her husband had been innocent; then she'd be happy over her release. Himmler told me: " The Fuehrer has accepted my proposition: I am now organising Department Seven of the Security Service: a Bureau for the Supervision of Reich Ministers." I was a little startled. Was this really necessary? Who might be entertaining thoughts of treason? Not Ribbentrop—that brave sheep admires the Fuehrer as much as before, not Goebbels, and the others are not important. But Hermann might be dangerous. Since that big quarrel with Adolf he has been sulking and perhaps he is plotting some mischief. But I don't want to ask questions; and no one is allowed to mention Goering's name in A.'s presence. But it is also possible that Himmlers wants to trap his great rival, Goebbels. I don't think he'll succeed. I don't think anybody knows except me what Goering said to the Fuehrer. It was a terrible moment when Hermann roared at him: " You are not normal, not normal, not normal! I know that you are important—but it is just in the important, abnormal things that the nature of your terrible secret is displayed!" I must forget it!

Spring, 1944

Adolf sent me a short story which he wrote with

his own hands and he added that it has a certain documentary value because it reproduces his own experience of some years ago almost word-for-word. When I saw that it was called "Bettina", I thought it might be the little Ribbentrop girl - but I was wrong. It's just a story some great passion of his - and his disappointment because she would not wait for him. Whether it's a good story I can't-judge; it was interesting for me - but why does he play about with such things now. of all times?

1944

Frau Elfriede K [Koch?] was arrested last week in a Morabit *hotel garni* during a raid of the vice squad. Her partner was a handsome paratrooper named Mitterwuzer. A bit unpleasant for the wife of the Gauleiter. *Obergruppenfuehrer* Hermann says that he roared with laughter when he read the report. She hadn't tried in any way to spare her husband but said, "Do you think that pig doesn't amuse himself in the east where they sterilize all those Ukrainian women so that they can have all the pleasure they want without any consequences?" But Frau Elfriede was treated like the others and passed through all the controls before they discovered who she was, whereupon she was released with a thousand apologies.

1944

Found a secret report on his table and read it! In the week of the 18th to 25th there were 2,726 executions inside the Reich, excluding the General government and the Czech Protectorate. Among them 219 women. The youngest was sixteen, the oldest sixty-three.

1944

The notorious and fearsome Himmler I know very well now. He visits me often and he doesn't always try like Goebbels to worm something new out of me about Adolf. I believe he knows enough in any case. Otherwise he is the most curious man I have ever met. –But as far as A is concerned, perfectly discreet. What's so funny about he has nothing to do. Not more than one file must remain on his desk at a time. Then He loves gadgets: he presses a button and the wall opens behind you and the guard becomes visible. He has built up a whole collection of complicated machinery; direct telephone lines to Rome, Budapest, Oslo and Stockholm, even to Switzerland; loudspeakers and listening apparatus. He says there is no official in the Chancellery or in the Ministries whose private conversations as well as telephone conversations he cannot overhear if he wishes to. "Shall we try?" he asked me several times. But the conversations of his officials do not interest me. He always carries at least three pincnez in case he loses or beaks one; also, two watches. He is the most punctual man I know. He needs two watches to control one with other. Only then has he the feeling that he always knows the exact time. He writes down everything he is told; he must have dozens of notebooks and A says this constant jotting-down makes him nervous, as if Himmler had no memory at all. He always writes with a tiny pencil which he produces from his trouser-pocket. And yet he has fantastic memory. He can always produce any personal data promptly He knows everybody's escapades, and about the people he hates, like Ley or Goebbels, he knows every tiny detail. He was the first person to know that Goebbels was running after the beautiful Hungarian wife of Anfuzo with his tongue hanging out. Of Anfuzo himself Himmler spoke with great sympathy. He was the first Italian, he said, whom he liked and who impressed him. For his reason alone he would prevent all intimacy between Goebbels and Frau A – "And even if my people have to force their way into the bedroom with guns in their hands, he wont get her," he said and swept his hand across the polished, empty top of his desk. It is really amazing: while Adolf suffers more and more from the pressure of war and has completely changed – he is nervous, irritable, full of depression and gloomy uneasiness – nothing is noticeable in Himmler. He is self-assured, completely calm, confident, cool and unbalanced. I believe he thinks more of his power, growing every month, his rising influence and his improved standing with fuhrer than of the war and how everything is going to end. He always says that the new weapon will decide our victory. Whoever employs in first, will win the war. If we had it, the war would be over in a few weeks. If we only had it!

1944

I changed my flat three times in a night. Twice the house was in flames when I came up from the shelter. How Ley *[head of the party organization and leader of the German Labour Front*] has invited me to share his emergency home. He that there were already twenty million people in Germany who have lost everything. I was frightened by this figure but Ley, who, by the way, has grown very old, told me, "With Fuhrer leading us, we'll overcome all these difficulties." He is right. He took me into his bedroom and showed me something which he said he has never shown anybody. On the chest of drawers there was an altar; in its centre was the Fuhrer's picture. In front of the picture there burned an "eternal light" – oil in a small red cup; right and left fresh flowers. "I do all this and evening I say my prayers here in front of 'his' picture and find great privilege of having a personal contact with him, the greatest of all, every day. " I was deeply impressed. Ley seems to surpass even me in his love for him.

Munich, 1944

At Giesler's [*Gauleiter of Munich*] they produced a thirteen-year-old Hitler youth today who has learned *Mein Kampt* by heart from beginning to end. We tested him thoroughly; we gave him any line and he continued without hesitation until we told him to stop. He really knows the whole long book by heart. We also asked him whether he had learned anything else by heart, poems, etc. He replied no, he wasn't interested in poems and he hadn't learnt anything else at all. Esser said this boy could appear in a circus later on and earn his bread – provided we win the war which has now become dubious. We all looked at him with gloomy faces. E sometimes has attacks of madness. On the way home E, who was very angry with the whole "production", told me, "A Hitler youth who knows *Mein Kampf* by heart but has never heard of Goethe is an absurdity in Germany. If we want this … we'll produce a generation of parrots but not men of independent thought, with bold spirits and free ideas. "Perhaps he is right, but it is nice to think and know that such a fresh, pleasant boy has learned the whole thick book by heart…

Summer 1944

I was at the GHQ when the news of the beginning of the invasion arrived. At first I was terribly frightened, but then I noticed that they all seemed relieved, as if some terrible pressure had been removed. At last we know where they are landing, said Colonel Muell. A sent me home at once. He really wanted me to go to Switzerland or Sweden. I must have looked at him quite aghast because he only shook his head impatiently and said: "Of course, only for some time, not for ever." But I don't want to go to Switzerland where they hate us Germans, nor to Sweden where they can hardly stand us. I want to stay in Germany and stick it out, like *he,* bravely, whatever may happen. My heart is full of fear. On the train the people were grumbling and some officers said to each other: "We have had enough, more than enough, peace must be made." We had to leave the train seven times because of air attack; the last time there were two people killed, a woman and a child. I stood there and looked at them without any fear: One gets hard. And we are expected to make peace!

Sunday, 1944

M told me today that the workers in the Reich call our retreat in the East the Emperor Napoleon Memorial Race. M thought this a brilliant joke. But where is the joke? Our soldiers are perishing out there, I told him, and you make such remarks – you'll end up on the gallows. I don't believe that our workers say such things. M went pale – the warning went that moment, and he took his leave much too hurriedly. I'll tell Himmler, such people must be hanged, they can't run around free and unpunished.

GHQ 1944

Today the Fuhrer told me with a dead-serious face, "I only have there people left whom I can trust: you, Goebbels and Himmler. You are the most faithful and you'll get your reward one day, I give you my solemn promise. Goebbels stands and falls with me. He has been gripped by the passion of the fight just like myself. He has grown above his own stature and reached one I would never have expected. Himmler is a chameleon. I warn you against him – I know you see him often. He is dangerous. He might become a danger even for me if he had any gift as an orator. But luckily he is dumb and practically sentenced to silence. Thus he can serve only me. But my generals are traitors and sentimental fools; they want to spare the troops. As if it weren't the most exacting and unsparing effort that always brought the biggest success. If I had any replacements, I would have three-fourths of the German generals shot. After the war they won't get the estates in the east for which they've been waiting so long – they'll be exiled or executed."More than ever *he* suffers from insomnia, I find him changed, his slim figure has acquired something buffalo-like and he walks with a little stoop, as if the burden of the whole of Germany rested upon him. Every evening he suffers terrible neuralgic pains which must be almost unbearable. These depressions, occasionally interrupted by brief exaltation and terrible excitement, show me that he is at the end of his tether. I also discovered that he can no longer make love – I believe he himself was the most deeply shattered by it. He probably thinks I haven't noticed anything. In such things he is so touchingly naïve. What woman wouldn't notice such a thing? But of course I left him with his faith undisturbed. He has now other and greater things to think about than me. I hope that he'll quickly forget the sad and humiliating experience of last night. In any case, it is probably a passing phase – I don't know…perhaps there is no such thing? I just figured out how old he is. It can only be a passing nervous weakness. But it shows that his psychological and physical state must really be causing anxiety…

GHQ 1944, Tuesday

Dr Morell came and I used the opportunity to ask him about the Führer's health. M Shrugged his shoulders and said that a different man would have collapsed long ago. But as *he* cannot be measured by normal standards, nothing could be said. I feel that he is evading my questions. He must feel that I, too, have changed; for he suddenly became serious and said, in shi best professional manner: "Physically there is nothing to be noticed except a certain additional strain. But he lives in a condition of mental exaltation that is sometimes accompanied by hallucinations. He also has a attacks of partial blindness, the cause of which I was unable to discover. That's the diagnosis and symptoms, my dear, in all sober truth. "My tears rolled down my cheeks. M isn't at all sentimental, he nudged me and said, "Come on, get hold of yourself, we can't have any weeping virgins here in GHQ." I shook myself and it was over. We must stick it out, the new weapons are coming and then God have mercy upon our enemies!

Summer 1944

I always think: perhaps you will see him for the last time today He won't be killed, I am sure, providence must hold its protecting hand over him. What would happen to Germany now if he died? That's unthinkable. But with me it's different. I am not afraid, I would never go into a shelter but defy the bombs. But *he* gave orders and so I obey. "The bombs don't inquire whether you are afraid or not," he said and as usual he was right. Just now we went for a walk in the summer fields and lay in the grass. Yes, there is no doubt anymore - out love has become platonic, I must accept the fact. I read about it, too. Perhaps after years of peace and relaxation it will be different again. But who thinks of such things now when everything is at stake?

1944, Wednesday

Met our English broad caster, Lord Haw Haw, at the Foreign Press Club now housed in Ribbentrop's villa. At last I had a few good laughs again n- not over anti-Nazi jokes which should really make one cry, but because I heard funny things about the Americans and the British. Haw-Haw is convinced that we'll win the war. His fresh humour is really an inspiration. He told me that London was terribly battered, much more than Berlin, only the British do not admit it. He is very well-dressed which makes him conspicuous in the general shabbiness. You hardly see a decent suit in the street. Later we discussed art; HH talked hair raising nonsense. It is so funny! The British probably have the best taste in men's fashions. The more amazing their complete lack of aesthetic culture in other directions…

1944

Dr M spent a day here and visited me. My first question was: "How is the Führer?" M, the cynical faun, hesitated, then he looked at me and said suddenly, "I asked him whether he had ever suffered from venereal disease. He grinned - for he cannot laugh, not for a year now – and replied. 'The usual childhood trouble, Nothing special.' I told him about the latest discoveries in this field; especially about the lecture of a very famous French specialist who had declared that syphilis usually kills its victims between the age of forty–five and fifty-five, at the stage when virility usually begins to ebb. You know what he replied? 'You ought to know my dislike of all that is French well enough to spare me such stupidities. The French, first of all, have a great inclination for melodrama which they cannot forget even in science – and secondly a tremendous talent for exaggeration. Their science, too, is exaggerated.' "It was terrible for me to hear this. And M watched me with a burning curiosity as if he wanted to read God-knows-what in my eyes. I asked myself what the significance of this tale was enquired, "What did you want to tell me by this?" "I only wanted to explain how ill the Fuhrer is," he said, "but also that there is no danger as long as he answers such concise warnings with political formulas." But I think good old M simply wanted to see how I would react to his revelation. What he told me was really incredibly swinish and a terrible breach of confidence. And so it is all along the line.

In his Reichstag speech on 26th April 1942, Hitler declared: "The decision will fall in the east." At the same time Goebbels was poking fun in "Das Reich" at the delay in the Anglo-American invasion. Since 1943, Supreme Command of Germany had known beyond doubt that the Western invasion was coming. Berlin, however, was putting out feelers about a possible separate peace with the Soviet Union. The conversation with Axmann, the leader of German Youth, may provide a key to Hitler's hopes for such an agreement once the Red Army had recaptured all the lost Russian territories. Perhaps Hitler was thinking of a development of the war in the East similar to that of the Napoleonic campaign in Russia.

Soon there won't be a single person who hasn't lost dear relatives and precious property in this war. If we do not achieve victory, we will become the poorest and most downtrodden people in the world and it wouldn't be worth living. Esser told me it didn't matter whether we won on the battlefield or not – morally we have won the war in any case. After the fantastic achievements of the German people no one would dare to say we were not the bravest and toughest – the ones who struck the hardest blows and could also take the most grueling reverses, E said. But a sad faced woman on the tram put it differently: "What's the use of final victory to me? I have lost my father, mother, sisters, husband and children. My house is in ruins. And in the end I might be sent to the east somewhere so that I'll even lose my country!" I believe the German people are terribly tired.

July 1944, Sunday

Met Axmann, who is on his way to Vienna from GQ. In Vienna, a traitor-group of the Hitler Youth with over twelve hundred members has been discovered. Their aim was "the removal of Adolf Hitler". It must have been a group with a strong religious basis for they called *him* "the Great Whore of Babylon" and "The Anti-Christ". They are said to have had local branches in the Rhineland which haven't been discovered yet because the head of the secret league, a young priest, died in the hands of the SS without confessing. Kaltenbrunner only carried out twenty executions because the boys are all under fourteen. Now A will try to clear up the matter from the inside. Axmann also told me that the Fuhrer asked him about his impressions of the Eastern front. "I told him: My Führer, the German soldier is the best in the world but there are too few of them. We need more divisions." And he roared at me: "And even if the Russians reach the Vistula, my best divisions stay in the west!"

23rd July 1944

Went for a long walk, knew nothing of what had happened and when I came home, found *his* cable: "I am alive and the war goes on!" At first, I didn't understand the whole things – but felt that something must have happened. I switched on the radio but there was only military music. Then E called me up and asked me to come at once – but not for him, to a friendly family whose name he gave me in code. I realized then that something terrible must have happened, packed a small suitcase with my jewelry and money and rushed off on foot because I didn't dare to use the tramway. Everything was quite normal in the streets, people looked much as usual and seemed to know nothing. The attempted assassination was still not known about. At the Zs, E told me everything he knew and it wasn't much. The Gauleitung was given direct orders by the Führer immediately to arrest all the military commanders of Munich with the help of the SS and the police. Then we heard the various orders on the radio and we realized that there had been almost a disaster at GHQ. It remains to be seen how widespread the conspiracy was. Giesler maintains Goering is involved, also Rommel, Rundstedt and I don't know who else. But the Luftwaffe is said to be loyal to the Führer, also the Navy and the SS. Kahrs just rang and said Berlin was fully under control and Goebbels has the power to guarantee order and security. Himmler has arrived in Berlin and has special powers – perhaps he'll become Commander-in-Chief of the army. Keitel is said to have been sacked and arrested with his staff. Search is being made for Stauffenberg's family, but they seem to have disappeared without a trace. Our friends all look pale and peaked. They all felt the breath of death this time. What would have happened to us if the attempt had succeeded? Who knows – perhaps we would all have been murdered. G said it was organized by the Intelligence Service;

for him it is nothing but proof of the weakness of the other side. If they must use such means, they must be in a bad fix. Probably they have given up now. He would be willing to bet, he said, that within four weeks there will be an Anglo-American peace offer. This was the British attempt to win the war. After this failure, they will be content with a compromise. But now, of course, there cannot be any compromise. G wanted to make bets about the peace offer – but there were no takers. After the excitement of this terrible day we were all dead-tired and yet could not go home. No one wanted to stay in his own place – they proffered the house of this nice Swedish family. The others, E and G, made themselves comfortable in this living room; I was given the bedroom of the daughter of the house who is in Stockholm at the moment. As I lay in bed, I took out *his* telegram and read it again and again. For the first time in ten years, perhaps even more, I tried to pray. But I was too tired to concentrate. I fell asleep with telegram in my hand – it must have been two or three in this morning.

The following four letters from Hitler addressed to Eva Braun Were affixed to the diary:

31.10.41

Maedi,

The magic of your demureness and of your eventual yielding bewitched me once again and I must tell your about it! You are my best comrade and no one has given me such joy as you. There is nothing in the world that can sublimate my wild greed for conquest except you.

Adolf.

5.5.1942

My love

I thank you for what you told me yesterday. I want to explain why I received it with less unequivocal joy than you had the right to expect. For you it is a simple, natural fulfillment; for me it means a tremendous, immeasurable responsibility. I'll bear this responsibility more than ever before. Do not forget that you have acquired various duties through it; greater perhaps than any task set for any women in Europe at this time. Will you remain conscious of this? Do not answer – but do not forget it for a single moment.

Adolf.

My dear,

You want a reply to your letters which arrive almost daily. Be not so impatient and do not try to threaten me with stopping to write to me! I know that it is a necessity for you to tell me your worries and you must know that it is an equal necessity for me to know them. Do not expect me, however, to assume that you could share my worries as I share yours! I think not only of you but of a whole nation and a whole continent, if not of the whole world. When I am with you, I must not forget it, otherwise the very earth will start to shake under the feet of millions of people. I have no objections if you apply childish conceptions to me. But I must tell you again and again: you must not insist that I should reply to you in the same terms. You ask me whether I-am well. In the sphere which providence has allotted for my daily tasks, men are neither well nor ill. I-am! For I am an absolute idea!

A.

Fifteen months later, three days before the July conspiracy, comes this cry for help, prompted by the despair of a man pursued by mad fears and hallucination:

Maedi,

I want you to fly here tomorrow! There are strange ideas in my brain. Sometimes, at night, I am overwhelmed by a mysterious anxiety which oppresses me. Come, come quickly!

A.
DESTROY THIS AT ONCE!

Printed in Dunstable, United Kingdom

91219618R00107